You Have Been Told What Is Good

You Have Been Told What Is Good

Interreligious Dialogue and Climate Change

PAUL O. INGRAM

CASCADE *Books* · Eugene, Oregon

YOU HAVE BEEN TOLD WHAT IS GOOD
Interreligious Dialogue and Climate Change

Cascade Books
An Imprint of Wipf and Stock Publishers
199 W. 8th Ave., Suite 3
Eugene, OR 97401

www.wipfandstock.com

ISBN 13: 978-1-4982-9348-8 (paperback)
ISBN 13: 978-1-4982-9348-8 (hardcover)
ISBN 13: 978-1-4982-9349-5 (ebook)

Cataloging-in-Publication data:

Names: Ingram, Paul O., 1939–.

Title: You have been told what is good : interreligious dialogue and climate change / Paul O. Ingram.

Description: Eugene, OR: Cascade Books. | Includes bibliographical references.

Identifiers: ISBN 13: 978-1-4982-9348-8 (paperback). | ISBN 13: 978-1-4982-9348-8 (hardcover). | ISBN 13: 978-1-4982-9349-5 (ebook).

Subjects: LCSH: Ecology—Religious aspects. | Climatic changes. | Human ecology—Religious aspects—Christianity. | Christianity and justice. | Islam—Relations. | Buddhism—Social aspsects. | Environmentalism—Asia—Philosophy. | Title.

Classification: BL65 E36 I53 2016 (print) | BL65 E36 (ebook)

Manufactured in the USA

In Memoriam

Gail Owens Ingram (1917–1999)
Lucille Ingram (1916–1981)
John William Ingram (1941–2003)
Robert Ray Inslee (1910–2006)
Ruth Davis Inslee (1909–1979)

Contents

Acknowledgments

ANYONE PRACTICING THEOLOGICAL REFLECTION and history of religions is in for a very difficult struggle. Like Jacob wrestling with God at the River Jabbok, we can't lose even as we are left with scars as we walk away limping for the rest of our lives. As theologians we search for perfect words knowing that we will always fail. This is so because as the mystics of all religious Ways instruct us, a God that is talked about, defined, or shrouded in liturgical practices or doctrinal abstractions is not God. God can't be reduced to our ideas about God. And yet we keep writing and speculating as we place our ideas and experiences of God into a theological system while knowing that everything we write or say will have to be revised and ultimately "unsaid." So of all professional academics, theologians have the most difficult task, even more difficult than physicists seeking to describe the processes at play in the universe or biologists unlocking the secrets of DNA.

For one trained history of religions—my field—consciously wearing the methodological hat of a process theologian makes life even harder. For I am wrestling with normative theological issues and descriptive accounts of the teachings and practices of several religious Ways in order to draw theological conclusions that I hope will initiate a planetwide resistance movement against the political and economic causes of climate change. The practice of interreligious dialogue always entails theological and historical issues; and much of this book seeks to place in dialogue the concepts of compassion, justice, and community in the Jewish, Christian, Islamic, Hindu, Buddhist, and Daoist Ways. In the process, the differences and congruencies between these religious Ways may serve as the foundation for this book's overall thesis: a socially engaged dialogue between the world's religious Ways focused on the meaning of compassionate, just community working for the common good of all human beings and sentient life forms might evolve into a powerful means for confronting the economic and political causes of climate change now placing the entire planet in imminent danger.

Even though writing is ultimately a lonely task of placing one's ideas into words that express what a writer has in mind, in point of fact no one ever writes anything alone. We rely on the insights of friends and colleagues in numerous fields of inquiry. We also rely on their critiques and criticisms of our ideas. We owe such persons our gratitude. One friend and colleague from Pacific Lutheran University has been particularly helpful. Professor Eli Berniker, who taught in the School of Business until his retirement, read chapter 7, "The Invisible Hand." The first part of this chapter is about the causal role of free-market capitalism in climate change. I am not an economist, but Eli understands the discipline of economics better than most economists I know. His sharp and precise critique of this chapter forced me to revise several sections. For this I am very grateful to Eli, a wonderful critic and friend.

But Eli Berniker is only one of a number of examples of a lesson my father taught me on a back packing trip more years ago than I wish to remember: no one makes anything alone. Beside standing on the shoulders of colleagues, friends, and teacher-scholars in the academic fields of history of religions and theology, I am also deeply grateful to the professionals at Wipf and Stock Publishers for bringing this book to final publication. In particular, I want to thank: K. C. Hanson, editor in chief, and his associates: Jeremy Funk, Heather Carraher, and Brian Palmer. I have worked with these professionals on other publications, and it is always a learning process I enjoy completely.

Finally, this book is dedicated to my brother, Richard Gail Ingram, whose sense of humor, encouragement, and dedication to his family has been so very inspirational in my life. My only regret is that we live twelve hundred miles apart.

Paul O. Ingram
Mukilteo, Washington

Introduction

IN AN ESSAY PUBLISHED in *YaleGlobal Online*, climate scientist James Hanson describes the running scam of the fossil fuel industry: overstating fossil fuel reserves.[1] The United States Department of Energy parrots this scam. The reality of increasingly limited supplies forces prices higher as fossil fuel corporations make more profits than ever in the history of the industry. They do so by continually asserting that plenty of oil, gas, and coal exist to be found, which has the effect of keeping consumers on their hook. In all probability, they may find more fossil fuels in deep oceans, under national parks, offshore, in the Arctic, and in other environmentally sensitive areas of the planet. But they do not require much actual oil, gas, or coal resources to keep people paying higher and higher prices.

In fact, the existence of fossil fuel reserves is a corporate illusion. "Coal reserves" are based on estimates made decades ago, and extractable coal reserves are overstated, which is consistent with present production difficulties and rising prices. The present estimate of a two-hundred-year supply of coal in the United States is a fiction. But the price of coal continues to rise and line the pockets of the coal industry's CEOs. In reality, conventional fossil fuel supplies are also limited even if the earth is torn up to abstract every remaining drop of oil and last shard of coal.

So if humanity is to live on a planet similar to the one on which civilization developed and to which all living beings on planet Earth have adapted, the rising CO_2 levels resulting from humanity's addiction to fossil fuels must be overcome by reducing CO_2 from its present level of 400 ppm (parts per million) to at most 350 ppm. It is possible that peak CO_2 can be kept to about 425 ppm if the current estimates of oil and gas reserves are true and if coal use can be phased out by 2030. Peak CO_2 can be kept close to 400 ppm if actual reserves are close to the estimates of those who believe

---

1. Hansen, "Timeline for Irreversible Climate Change."

that the planet is already at peak oil production, having extracted half of existing fossil fuel resources.

But it is clear that a different path must be followed, one reflective of our obligation to be stewards of the earth's natural resources in a way that benefits all sentient beings. This different path is premised on the possibility of achieving 350 ppm CO_2 levels or lower this century, provided we move beyond fossil fuels soon because a large fraction of CO_2 emissions will linger in the atmosphere for many centuries. But choosing this path will require headlong confrontation with the economic forces driving free-market capitalism and our unbridled consumerism of the earth's natural resources. Since rising CO_2 levels are the major source of rising temperatures, melting ice caps, destructive weather events, coastal flooding and other environmental disasters, resistance to the economic and political domination systems profiting from oil, gas, and coal production must be global. The issues are too important and dangerous for any one nation to resolve. Hence the thesis of this book: an international, socially engaged interreligious dialogue focused on climate change and the resulting environmental damage with its accompanying cultural disasters might be a force to reckon with in seeking the means of living responsibly on what may be the only planet in the universe capable of supporting life.

In point of fact, an international, socially engaged interreligious dialogue on climate change may already be in the making. In a report published in London in the *Daily Mail* a number of scientists, inspired by Pope Francis's encyclical *Laudato Si*, have concluded that the voices of faithful religious persons representing all of humanity's religious Ways need to be brought into the conversation about confronting not only the physical causes of climate change but also policy decisions about how to combat it.[2] Increasingly, people are coming to the conclusion that religious faith is able to mobilize broader movements of human beings for action on climate change than the scientific community by itself. Or as I interpret the *Daily Mail*'s report, a global, socially engaged interreligious dialogue that includes scientists as a third partner is our best hope for confronting the causes and combating the negative effects of climate change. The issues are indeed global. What is needed is a global response.

The title of this book, *You Have Been Told What is Good: Interreligious Dialogue and Climate Change*, seeks to bring together the academic disciplines of history of religions, theological reflection, interreligious dialogue,

2. Borenstein, "Scientists Enlist the Big Gun."

and dialogue between science and religion to bear on the issues of climate change. Normally, these disciplines are specialized into academic subdisciplines with intellectual boundaries so tightly drawn around their particular methodological borders that specialists in these fields mostly talk to one another in their technical language while ignoring scholars in other academic disciplines isolated in their particular technical jargon. So most scholars speak their own language to one another on highly technical and specialized topics while ignoring problems requiring more interdisciplinary approaches. Scholars hemmed in by their balkanized academic boundaries simply cannot address problems of human survival. Climate change caused by global warming is perhaps now the most important of these problems. However, interreligious dialogue is by its very nature interdisciplinary. Because I think this is so, socially engaged interreligious dialogue focused on climate change, involving all persons of faith and those working in the natural sciences, is important beyond any measure I can conceive.

The title of this book comes from my translation of Micah 6:8: "You have been told, O Mortal, what is good, and what God requires of you: do justice, do compassion, and walk with God in solidarity of community." The radical interdependency of justice, compassion, and solidarity of community working for the common good are ideals celebrated in the religious Ways of humanity within the pluralist contexts of each Way's distinctive worldviews, teachings, and practices. Human beings at all times and in all places have known what is good, but for reasons too numerous to count have failed to act justly and compassionately in communal harmony with one another and with the sentient beings with whom we share life on planet Earth. Today, the major justice issue confronting us is human-caused environmental destruction running amok on this planet, the only place in the universe where our species is alive.

Accordingly, the first three chapters of this book are concerned with the meanings and interdependency of justice, compassion, and solidarity of community in five religious Ways: Jewish, Christian, Islamic, Hindu, Buddhist, and the Chinese Ways—meaning Daoist and Confucian traditions. The title of chapter 1 is "Justice." In the Hebraic prophetic writings preserved in the Tanak, the meaning and practice of justice and its relation to compassion and solidarity of community are spelled out with stunning clarity. "Justice," in Hebrew *mišpaṭ*, means "right treatment of human beings," which prophets like Micah described as giving to human beings what they need for meaningful existence. After spelling out the nuances of the meaning of justice for prophets like Micah, this chapter continues to explore the meanings of justice in the religious Ways mentioned in the preceding paragraph as evidence for this chapter's thesis: in spite of the differences between humanity's religious Ways, the nature of justice is strikingly congruent in the Ways surveyed here, with the possible exception of the Buddhist Way.

Chapter 2, "Compassion," seeks to clarify not only the meaning of "compassion" but also the interdependency of justice and compassion. "Compassion," in Hebrew *ḥesed*, means experiencing another person's or community's suffering or joy as one own and relating to that person or community justly. For Hebraic prophets like Micah, compassion and justice are so interdependent that one cannot exist without the other. That is, according to the religious Ways summarized in chapters 1 and 2, justice and compassion can never be separated because justice without compassion is a tyranny, while compassion without justice is mere sentimentality. The interdependency of justice and compassion resembles the interdependency of *yin* and *yang* in the Chinese Way. Justice and compassion meet at that exquisite point of balance between extremes that the Hebrew prophets called "solidarity of community."

"Solidarity of community," in Hebrew *ṣedaqah*, is the topic of chapter 3. According to the Hebraic prophets as well as the religious Ways summarized in this book, the foundation of community is justice and compassion. Consequently, speaking or writing about solidarity of community is impossible in isolation from justice and compassion. Community is where justice and compassion meet for the common good of human beings in interdependency with the common good for all sentient beings and the environmental processes supporting all life on planet Earth.

Chapters 1–3 are the foundations for the remaining chapters. In chapter 4, "The Invisible Hand," I analyze capitalist economic theory and contemporary consumerism as the primary force driving the current environmental crisis. Contemporary interpretations of capitalist economic theory portray human beings as isolated individuals separate from other individuals, seeking their own good in indifference to the successes or failures of others engaged in the same activity: this is an example of the sort of abstraction Alfred North Whitehead called "the fallacy of misplaced concreteness" because of the way most economists abstract from the actual communal character of human existence. Yet in the real world, the biological evidence is in: *self-contained individuals do not exist*. Thus the individualism of current economic theory is profoundly erroneous because human beings are social, that is, are constituted by their relationships to other human beings *and* to the life forms with which we share planet Earth. That human beings are constituted by interdependence, not separation, with all things and events is clearly a central teaching of the Buddhist Way, but it is also underlies Israelite and Judahite prophetic tradition, the life and teachings of the historical Jesus, Islam's call for compassionate justice, the Hindu Way, the Chinese Way, and the Jewish and Christian Ways.

"Interreligious Dialogue and Resistance against Ecological Injustice" is the title of chapter 5. Human beings have struggled to create just, compassionate communities throughout history. Against all odds this struggle continues in our own time. Humanity's religious Ways everywhere took the lead in this struggle, more often than not against conventionally religious human beings and institutions. This struggle remains ongoing, most often against great odds in opposition to powerful economic and political domination systems along with the domination systems that inhabit all religious Ways: powerful bishops, priests, monks, imams, and rabbis placing their collective political and economic interests ahead of serving the common good.

So the call to live justly in compassionate community has come from a minority of religious voices: the entire Mosaic tradition celebrated by Jews as interpreted through the oracles of the eighth-, seventh-, and sixth-century-BCE prophets; the historical Jesus, whom Christians confess as the Christ of faith; Mohammed; Gautama the Buddha; the Daoist Sages, and Confucius. Contemporary examples abound as well: every rabbi I've ever met, Martin Luther King Jr. and Thomas Merton, Badsha Khan and Mahatma Gandhi, Wing-tsit Chan, and lay followers of each religious Way. Clearly, resources for resisting the economic and political systems responsible for global environmental injustice are available in humanity's religious Ways, but only if some means can be found to unite faithful persons inhabiting these Ways in a more unified struggle for the common good that includes all sentient beings now inhabiting planet Earth.

Chapter 6, "The *Praxis* of Interreligious Dialogue," is about the interrelation between the practice of dialogue and social engagement with issues of environmental injustice. The evidence seems clear: the world's religious Ways have instructed us about what the common good requires through an incredible plurality of teachings and practices. Or restated somewhat differently, from this plurality emerges a profound thirst for compassionate, just communal structures working for the common good. Given this worldwide search, interreligious dialogue must focus on: (1) the common meanings of compassion, justice, and community within each religious Way; (2) the different nuances of meanings regarding compassion, justice, and community; (3) the meaning of "the common good" within each religious Way; and (4) how to create political, social, and economic structures for compassionate communities that reflect humanity's "unity in diversity."

It is important, therefore, that socially engaged interreligious dialogue be grounded in conceptual and interior dialogue. Understanding the teachings of religious Ways other than our own and participating in their distinctive practices—prayer, meditation, communal activities like feast days and celebrations of important events—deepens one's understanding of one's own faith. Yet it is also the case that some persons find it necessary to leave the home of their original faith community and enter the faith community of their dialogical partner. The practice of interreligious dialogue is not for the fainthearted. Either way this process goes, we become less parochial as we become open to religious diversity.

Of course there is a good deal of fundamentalist nonsense in all religious Ways. Fundamentalism is a dangerous distortion of the Christian Way. Radical Islamic groups are not groups that follow the Qur'an's injunctions against terrorism and the oppression of women. Current Jewish oppression of Palestinians by the government of Israel does not reflect the Torah's or the prophetic injunctions to compassionately and justly struggle for the common good of all. Violence, racism, war, or the oppression of the poor by the rich justified by any religious Way is a distortion of religious faith and practice whenever and wherever it occurs. Interreligious dialogue exposes both the distortions of religious faith as well as the creatively transforming teachings and practices of humanity's religious Ways. So dialogue entails becoming "wise as serpents." Speaking for myself, the more I have known faithful persons living at the depths of their distinctive religious Ways, the more my own particular Lutheran faith has been deepened, stretched, and creatively transformed.

The topic and title of chapter 7 is a qustion: "What Can Be Done?" *Community* is an overused word. Economists and politicians on the Left and Right use it the way fast-food restaurants use salted fat to cover up the lack of healthy ingredients in their food. Religious persons are often just as ambiguous regarding just what constitutes a community. But in the prophetic literature of the Tanak, the meaning of community is inclusively clear. This is so in the teachings of the historical Jesus whom Christians confess to be the Christ of faith, as well as in the teachings of Islam, in the Hindu Way, in the Buddhist Way, and in the Daoist and Confucian Ways as well. As opposed to social clubs and mere collections of human beings living in the same neighborhoods, *community* names an inclusively compassionate just social-political structure that strives for the common good of all. We need to rescue this meaning of *community* so that it becomes the most clearly understood word in our thinking in this contemporary time of religious pluralism and dangerous global warming with its resulting environmental disasters.

So we have been told what is good throughout human history beyond human counting. But here's the problem. Access to endless amounts of cheap energy for those of us wealthy and lucky enough to live in an industrialized country has made us rich in comparison to the majority of the poor who live unindustrialized countries. Capitalist economics treats all human beings as independently separate from other persons, the "rugged individual" of an Enlightenment ideal going back to Descartes. Our present economic system is designed to work without input from our neighbors next door or down the block. Because of cheap oil, our food arrives from great distances with little inconvenience to ourselves other than shopping for it at the local grocery store. Credit cards and Internet connections make it possible to order much of what we need or desire and have it left anonymously on our doorsteps. In short, we've adapted to a neighborless existence and the near disappearance of community.

It is obvious that capitalist economic theory and practice is the engine driving the corporate greed for increasing the profits of stockholders, while consumerism is the fundamentalist faith of most individuals on this planet. The desire for more and more stuff runs wild even among those who identify themselves as Jews, Christians, Muslims, Hindus, Sikhs, Buddhists, Daoists, or Confucians. But in point of fact, consumerism is contrary to the religious Ways of humanity. Based on the assumption that human beings are individuals-in-separation, the interdependent relationships foundational to the creation of just, compassionate communities working for the common good have little chance of evolving. Yet we have been instructed about what the common good requires by a plurality of teachings and practices in humanity's religious Ways. From this plurality emerges a profound thirst for compassionate, just communal structures working for the common good. It is this collective thirst, this global thirst, that must be transformed into willingness to resist the economic and political domination systems eating this planet alive, before its to late.

1

The Way of Compassion

THE JEWISH WAY

More or less seven hundred years before the historical Jesus of Nazareth was executed for his radical public criticism of the oppressive Roman and Judean political, economic, and religious domination systems of his time, the prophet Micah railed an identical critique of the oppressive power structures of his day. Standing somewhere in or near Jerusalem, he shouted:

> With what shall I come before the LORD
> and bow myself before God on high?
> Shall I come before him with burnt offerings,
> with calves a year old?
> Will the LORD be pleased with thousand of rams,
> with ten thousand rivers of oil?
> Shall I give the firstborn for my transgression,
> the fruit of my body for the sin of my soul?
> He has told you, or mortal, what it good;
> and what the LORD requires of you
> but to do justice and to love kindness,
> and to walk in solidarity of community with God. (6:6–8)[1]

1. Verse 8 is my translation.

1

A few years before Micah, the prophet Amos standing somewhere in Samaria, the capital of the northern Israelite kingdom, shouted a similar message:

> I hate, I despise your festivals
>> and I take no pleasure in your solemn assemblies.
> Even though you offer me your burnt offerings and grain offerings,
>> I will not accept them;
> and the offerings of your fatted animals
>> I will not look upon.
> Take away from me the noise of your songs;
>> I will not listen to the melody of your harps.
> But let justice roll down like waters,
>> and righteousness [solidarity of community] like an ever-flowing stream. (Amos 5:21–26)

Each of the eighth-, seventh-, and sixth-century prophets whose oracles and sayings are preserved in the Tanak stood squarely within the Exodus traditions of God's covenant with Israel, a covenant requiring the creation of community established on justice and compassion so as to be a "light to the nations," as Deutero-Isaiah put it—an example of solidarity of community for other societies to imitate in their own particular ways. Everything else—religious obligations and practices, legal obligations, economic and institutional social hierarchies and power structures—are of less than secondary importance. In particular, religious teachings and practices must not be employed as an ideological cover-up to justify violence and injustice.

In Hebrew, the word translated into English as "justice" is *mišpaṭ* or "right treatment of human beings," which according to the biblical prophets depended on creating communal structures upon which human beings must depend for peaceful communities, which too often is *not* what human beings actually experience.

The nondual side of *mišpaṭ* is *ḥesed* or compassion. "Compassion" is the experience of another human being's or human community's joy or suffering as one's own and relating justly to that person or community accordingly, particularly through resisting personal and institutional forms of injustice. The foundation of compassion is the utter interdependence of human beings in community. Or to restate this in language appropriated from philosophical Daoism, justice and compassion are the nondual *yin* and *yang* that constitute the foundation of *ṣedaqah*, usually translated as "righteousness." But the English word "righteousness" does not adequately convey the nuances of *ṣedaqah*. *Ṣedaqah* will be the focus of chapter 3. For now, it is useful to note that a more accurate translation than "righteousness is "solidarity of community," which from the second century BCE until the present was imagined as a "kingdom of God" finally established by God with the appearance of the Messiah. The first-century Jesus movement affirmed the coming of a the future messiah, whom after his death Christians identified with the risen Jesus of Nazareth.[2] For the early Jesus movement the commonwealth of God is not only a reality fulfilled at a future time of God's choosing but is also a present reality whenever and wherever human beings resist domination systems that oppress other human beings.

As I noted, the English word for *hesed*, is "compassion." Literally, "compassion" means "to suffer with," and it occurs when one experiences the suffering of other human beings as one's own and is thereby motivated to relieve that suffering because it *is* one's own. The English noun "compassion" derives from Latin. Its first syllable comes directly from *com*, an archaic version of the Latin preposition and the affix *cum*, meaning "with." The "passion" of "compassion" is derived from *passus*, which is the past participle of the verbs *patior, patī, passus sum*. "Compassion" is thus related to the meaning of the English noun "patient" ("one who suffers"), from *patiens*, and is akin to the Greek verb *paskhein* ("to suffer") and to its cognate noun *pathos*.

But "compassion" does not merely name a subjective feeling in response to suffering, at least according to the Jewish Way. This is so because compassion is grounded in the utter interdependency of all things and events, apart from which there would be no things and events, because of God's continuing creation of the world. This means all human beings are "relatives," part of an extended family that some liberal rabbis have said includes all life forms with which we share planet Earth.

Accordingly, "compassion," ontologically grounded in interdependence, is what happens when human beings wake up to fact that we and all life forms dwell in a world in which no thing or event is separate from any other thing and event. In other words, we have never been, nor now are, separate beings. The suffering of others as well as the suffering of living beings with whom we share this planet—*is our own suffering*.

---

2. In Greek, *basileia tou theou*, translated as "kingdom of God" in the New Testament, but a more accurate translation is "commonwealth of God."

This is one of the reasons I have long thought that the story of Jacob's combat with God at the River Jabbok is the major paradigm encompassing the journey of faith for Jews, Christians, and Muslims.[3] Wrestling with God requires following God's "instructions" in the Torah to create compassionate, just communal structures of existence. It is more often that not a bruising experience. For wrestling with God requires faith—meaning "trust," not "belief in doctrines"—that requires betting one's life on something: in Jacob's case, on what he encountered in his wresting match with God. In Judaism, Christianity, and Islam, "faith" is God's way is starting a fight with us. A hip—or something else—will be thrown out of joint—and we will limp, like Jacob, throughout the remainder of our lives.

As the history of Jewish experience records, wrestling with God always leaves faithful Jews with scars because "being Jewish" means participating in a communal effort to establish compassion and justice as the heart of God's Torah ("Instructions"). It was for this that Jews have been "chosen," so as to be a "light to the nations," as the Deutero Isaiah put it (49:6). But it is one thing for a community to be *called* to live justly and compassionately; it's quite another thing to figure out how to create communal structures of existence that are compassionate and just. So in imitation of limping Jacob, whom God renamed Israel, meaning "he who wrestles with God and wins," Jewish experience is a never-ending history of wrestling with God to figure out *how* to live in accordance with the Torah's instructions throughout the ever-shifting conditions of Jewish history.

According to the Tanak, the foundational model for the practice of compassion is the feeling of a parent for a child, as exemplified in Ps 103:13: "As a father has compassion for his children, / so the LORD has compassion for those who fear him." But compassion should characterize the relations between all human beings because living compassionately is evidence that compassionate human beings are deserving recognition as "blessed by God" (see 1 Sam 23:21). And in Zech 8:9, compassion is included among the postulates of humane interactions between persons. Inversely, the lack of compassion marks individuals and nations as "cruel" (Jer 23). The poor are especially entitled to compassion, and the repeated requirements of the Torah and the prophetic oracles that the "widow," the "orphan," and the "stranger" shall be protected show how deeply rooted the practice of compassion is in Jewish tradition. As one reads, for example in Exodus, Deuteronomy, and Leviticus, or in the stories of Saul, David, or Solomon,

---

3. Gen 32:24–31. Also see Ingram, *Wrestling with God.*

or the prophetic oracles, the impression is that compassion should be the center of human relationships. Yet persons alive during the times described by the biblical narratives, as well as persons alive today, too often experience compassion as a goal that the majority of persons and communities choose to ignore.

Physiologically and psychologically, the Tanak places the seat of compassion in the bowels. But the eyes were credited with the function of expressing compassion. God is portrayed as "full of compassion" (Ps 103:11), and this compassion is promised to human beings, as in Lam 3:22: "The steadfast love [compassion] of the LORD never ceases, / and his mercies never come to an end." Finally, Yahweh is declared to love the poor, the widow, the orphan, and the stranger. God is therefore named "Gracious and full of compassion" (Exod 34:6). Furthermore, even non-Jews experience God's compassion, as Jonah discovered to his dismay.

THE CHRISTIAN WAY

The prophetic call for justice, compassion, and solidarity of community was also the heart of the teachings of the historical Jesus. This is not surprising, since he was a first-century Israelite whose teachings were grounded in the prophetic traditions preserved in his day. Accordingly, the New Testament's word, *agapē* or "love," to describe the historical Jesus's teaching is not the Hebrew word he would have employed. As a first-century Israelite, he used the word *ḥesed*. "Love" in biblical Hebrew (*'ăhavah*) has strong erotic connotations absent from the prophetic understanding of compassion at the center of the historical Jesus's teachings and actions.

Most Jesus scholars seem to be in agreement that Jesus was both a mystic and a political activist. Like most mystics, Jesus had both apophatic experiences of union with God (experiences indescribable with words, which cannot tell the fullness of God) and kataphatic experience of God's continual presence (experiences of God describable with words). As a teacher of wisdom—a "sage," as wisdom teachers are commonly called—and like the biblical prophets before him, he publicly criticized the unjust power structures of his day. Compassion and justice were the topics of all of his authentic aphorisms and parables preserved Synoptic Gospels.

But there are two types of wisdom, which means there are two types of sages. The most common wisdom is conventional wisdom, and its teachers are conventional sages. Conventional wisdom is "what everybody knows," a culture's understanding about what is real and how to live in accordance with what is real. The second type of wisdom is an alternative, subversive wisdom that undermines conventional wisdom and points to another path or "way of life." Its teachers are subversive sages. For example, the historical Buddha and the Daoist sage Zhuanzi taught a "Way" that leads away from conventional perceptions and values toward a way of life that reflects "the way things really are." The wisdom of subversive sages is the wisdom of "the road less traveled," whose basic character was described by the thirteenth century Christian beguine mystic Maguerite Porete as "living without a why."[4]

The historical Jesus spoke of "the narrow way" that leads to life and subverts "the broad way" followed by conventional human beings, which leads to injustice and death. But to understand the "narrow Way" of which Jesus spoke, it is necessary to consider what Jesus taught about "living without a why."[5] Understanding how the historical Jesus described the narrow way of subversive wisdom as the Way of justice and compassion should provide Christians a set of interpretive lenses through which to dialogically engage the subversive wisdom of non-Christian sages.

It is well established that Jesus, like the Israelite and Judahite prophets before him was an oral teacher who employed aphorisms and parables. Aphorisms are short, easy-to-remember sayings, like great one-liners. Parables are essentially short stories. Together, the aphorisms and parables preserved in the Synoptic Gospels place us directly in contact with the voice of the historical Jesus because according to contemporary Jesus scholarship the most certain thing we know about him, since he lived in a culture where literacy rates were quire low, was that he was a storyteller and speaker of great one-liners. The aphorisms and parables of Jesus are invitational forms of speech. Jesus used them to invite his hearers to apprehend something they might not have otherwise apprehended. Parables and aphorisms tease imagination into action, suggest more than they directly say, and invite a transformation of perception and understanding. In many ways, they function like koans in Zen Buddhist meditational practices.

Jesus's aphorisms, more than a hundred of which are recorded in the three Synoptic Gospels (Matthew, Mark, and Luke), are crystallizations that invite further reflection as they more often than not generate startling insight. "You cannot serve two masters"; "You cannot get grapes from a

---

4. Porete, *Mirror.*

5. See Ingram, *Living without a Why*, ch. 7.

bramble bush"; If a blind person leads a blind person, will they not both fall into a ditch?" "Leave the dead to bury the dead"; "You strain out a gnat and swallow a camel"—all are short provocative one-liners that say more than their literal meanings and invite hearers to apprehend something they otherwise might not understand.[6]

Jesus's aphorisms were probably spoken one at a time, but this is not how they appear in the Synoptic Gospels, where they are typically grouped into collections of sayings. Aphorisms are also said many times. No oral teacher, especially an itinerant teacher like Jesus, uses a one-liner only once. This means their particular context described in the gospel narratives was not the sole context in which they were heard. It is perhaps best to imagine Jesus's aphorisms as repeated pieces of oral teaching employed many times on different occasions.

Some of the parables are very short, as brief as a typical aphorism, with the only difference being that they are narratives. Jesus's short parables, like aphorisms, are memorable, enigmatic sayings complete in themselves. For example:

> To what should I compare the kingdom of God? It is like yeast that a woman took and mixed three measures of flour until all of it was leavened. (Luke 13:20 = Matt 13:33)

> The kingdom of heaven is like treasure hidden in a field, which someone found and hid; then in his joy he goes and sells all that he has and buys that field. (Matt 13:44)

But most of Jesus's recorded parables are similar to short stories with plot and character development. It is probable that Jesus would have told them numerous times in different ways and may have expanded them to different lengths depending on his audience.

Jesus employed aphorisms and parables to subvert conventional wisdom and replace it with subversive wisdom. Conventional wisdom is the dominant wisdom of any culture, a culture's most taken-for-granted understanding about the way things are and about the way to live in harmony with the way things are. Conventional wisdom, in other words, summarizes a culture's dominant worldview. Conventional wisdom is a culture's social construction of reality and the internalization of that construction within the psyche of individuals. These conventional social constructions of reality

---

6. Luke 16:13 = Matt 6:24; Luke 6:44 = Matt 7:16; Luke 6:39 = Matt 5:14; Luke 9:60 = Matt 8:22; Matt 23:24.

are opposed to the prophetic call for compassionate and just community. Therefore, conventional wisdom offers guidance about how to live and covers everything from highly practical issues such as etiquette to images of the good life.

Moreover, conventional wisdom is supported by institutionalized systems of rewards and punishments: you reap what you sow; follow this way and all will be well; you receive what you deserve; the righteous will prosper—these are examples of the constant messages of conventional wisdom. Finally, conventional wisdom has both social and psychological consequences. Socially, conventional wisdom creates a world of hierarchies and boundaries. Some of these are inherited, exemplified when differences in gender, race, or physical condition are given hierarchical values and roles. Psychologically, conventional wisdom becomes the basis for personal identity and self-esteem. Politically, conventional wisdom is the ideological foundation of oppressive social, political, and economic systems of injustice.

But the conflict between the ways of subversive and conventional wisdom is not only part of the histories of Judaism and Christianity. For the subversive wisdom of the world's religious Ways affirms compassion, justice, and solidarity of community in their own distinctive ways as means of resisting oppressive communal structures of existence. The struggle against unjust domination systems is ongoing and crosses religious boundaries and continues to this day. But there is great diversity in how various religious Ways have reflected on justice, compassion, and solidarity of community. But there are also great similarities. My claim is that the similarities and differences are points for entering a socially engaged interreligious dialogue on contemporary environmental issues as a means of establishing solidarity of community for the common good in an interdependent world.

But there is hard truth: resisting injustice is one thing; creating just communal systems free from unjust domination systems remains an unrealized goal in spite of human response to the prophetic call to resistance. Communal systems of justice and compassion remain a perpetual ideal to be pursued, but perhaps never fully realized—which does not mean that we should not take the road less traveled in the global struggle for just, compassionate community.

This is certainly the view of 2 Corinthians, where God is spoken of as the "Father of compassion" and the "God of all comfort":

Blessed be the God and Father of our Lord Jesus Christ, the Father of mercies and the God of all consolation, who consoles us in all our afflictions, so that we may be able to console those who are in any affliction with the consolation we ourselves are consoled by God. For just as the sufferings of Christ are abundant for us, so also our consolation is abundant through Christ. If we are afflicted, it is for your comfort and salvation; if we are comforted, it is for your consolation and salvation; if we are being consoled, it is for your consolation, which you experience when you patiently endure the same sufferings that we are also suffering. Our hope for you is unshaken, for we know that as you share in our sufferings, so also you share in our consolation." (2 Cor 1:3–7)

So for Christians, the historical Jesus declared to be the Christ of faith embodies the model of compassionate justice according to which human relationships are to be constructed. Compassion, as the parable of the Good Samaritan holds up, should extend to all human beings, including one's enemies.[7]

THE ISLAMIC WAY

"Compassion" is the most frequently occurring word in the Qur'an. Each of its 114 chapters, with the exception of the ninth chapter, begins with the invocation, "In the name of God, the Compassionate, the Merciful . . ." The Arabic word for "compassion" is *rahman*. *Namaz*, or daily prayer five times a day that is one of the Five Pillars by which the Qur'an defines the practice of *'islām* ("surrender" to God's will"), begins by invoking "Allah, the Merciful and Compassionate." "Compassion," *rahman*, is one of the "Ninety-nine Beautiful Names of God."

As in Jewish teaching and particularly Christian mystical tradition, *rahman* is what human beings experience whey they become aware of God's creation of the universe that conjoins together all things and events into a web of mutual interdependence. Thus another human being's suffering is, in part, one's own since all human beings are brothers and sisters. Compassion is a relationship in which one actively works to relieve suffering wherever it occurs in the human community in general, referred to by the Qur'an as the "House of Islam." At the heart of Islam is the principle of *tawhid*, or the "oneness" or "unity" of God. This "unity" is intrinsically

7. Luke 10:25–37.

linked to the concept of compassion. Thus, "My Mercy encompasses all things" (Qur'an 7:156). Accordingly, the Qur'an identifies sentiments like love and compassion as expressions of the interconnected unity of all human beings that reflects the oneness and unity of God.

The Qur'an frequently speaks of God's plan for diversity, and the goodness of diversity as an ingredient in God's and humanity's "unity." For example,

> O mankind, we have indeed created you as male and female, and made you as nations and tribes that you may come to know one another. (49:13)

> And every community has its direction of which He lets them turn towards it. Vie, therefore, with one another in doing good works. Wherever you may be, God will gather you all unto Him. (2:148)

> There is no compulsion in religion. (2:256)

> Truly, those who believe, and the Jews, and the Christians and Sabeans—whosoever believes in God and the Day of Judgment and acts virtuously will receive their reward from their Lord; no fear or grief will befall them. (2:62)

> Whoever saves the life of one human being, it shall be as if he had saved the whole of humankind. (5:32)

The great mystical writers of Islam, the Sufis, wrote constantly of love and compassion as essential to the practice of Islam. They described compassion as the remedy for all ills and the alchemy of existence; compassion transforms poverty into riches, war into peace, ignorance into knowledge, and hell into Paradise. For example, Jalal al-Din Rumi, who was born in modern-day Afghanistan in 1207 CE, is arguably the best known of all the great Sufi mystics. Rumi wrote that while compassion is the foundation of the Sufi Way, it is a reality that can only be known by experience in order to be truly understood. In his words: "Love [Compassion] cannot be contained within our speaking or listening. It is an ocean whose depths cannot be plumbed . . . Love [compassion] cannot be found in erudition and science, books and pages . . . The kernel of Love [Compassion] is a mystery that cannot be divulged."[8]

The Sufis in their own individual poetic languages understood that the compassionate character of love entails embracing diversity, or what we call today, "pluralism," rooted in the experience of the transcendence of self. Transcendence of self—that is, of the illusion that we exist in separation from other human beings, God, and the life forms with which we share the earth—is not the only source of Sufi spirituality. For example, Abu Bakr Muhammad Ibn Arabi, (1165–1240), who was born in the Andalusia region of Spain, is considered one of the greatest writers of the Islamic mystical tradition. Known as the poet of "the Cosmic Heart," he speaks to us of "Discovering the Deeper Grounds of Suffering in Opening the heart." He writes: "Thus the person who understands the meaning of suffering increases his loving-compassion for the one who is in pain will be rewarded . . . ," because "as the Arabic proverb expresses it, 'every moist heart is a divine reward.'"[9]

THE HINDU WAY

What non-Hindus name "Hinduism" is an abstraction from what Hindus who practice the Hindu Way actually do and experience. That is, the label Hinduism covers a religious tradition in which every idea, practice, and myth that has ever occurred in human imagination can be found in the some form in the Hindu Way. Of course pluralism is an ingredient of all humanity's religious Ways. Just how many ways are there of being Jewish? Or Christian? Or Muslim? But unlike the pluralism of Christian tradition, Jewish faith and practice, Islam, Buddhist tradition, or the Ways of Confucian and Daoist cumulative traditions, there exists no defining teaching or practice characterizing the pluralism of "Hinduism" to which all Hindus assent. This being said, compassion is a virtue with many shades of meaning in the classical texts of the Hindu Way, meanings that center on the word *darśan* or "seeing" or "apprehension." This can be illustrated by a memorable episode in Indian mythology.

Once upon a time, the god Śiva and the goddess Pārvati were sporting in their high Himalayan home when Pārvati covered Śiva's eyes with her

---

8. *Diwan-i-Shams-I Tabrizi,ī,* ("The Works of Shams of Tabriz"), cited in Schimmel, *Mystical Dimensions of Islam,* 134.

9. Austin, trans., *Bezels of Wisdom,* 103.

hand. The whole universe was plunged into darkness. For when Śiva's eyes are closed there is no light anywhere, except in the fire of his third eye, that always threatens the destruction of worlds. The all-seeing gods are said never to close their eyes, and from the near disaster of Śiva's and Pārvati's play, it is clearly a good thing they do not, because the well-being of the world is dependent upon the open eyes of the gods.

So according to the Indian Way, not only must the gods keep their eyes open, but so must we, in order to make contact with them, to reap their blessings, and to know their secrets. When Indians go to a temple, their eyes meet the powerful, eternal gaze of a deity. It is called *darśan*, "seeing" the gods as they really are and being seen by the gods as we really are, without delusion, or *māyā*. The practices and traditions of the Hindu Way are all attempts to see or apprehend Brahman—of which the numerous deities of Hindu faith and practice are limited "incarnations" or *avatars*—without delusion and so to live without delusion. Thus if there is a single thread uniting the Hindu Way, it is *darśan*. For reality, "the way things really are," can be seen everywhere at all times and in all places—if we know how to look. The Hindu Way is a pluralism of ways of training human beings how to "see" or "apprehend" this "reality," usually named "Brahman" or "Sacred Power" that is "in, with, and under" the particular things and events of existence.

Accordingly, one can understand that the plurality of Hindu teachings and practices are methods of training persons to apprehend by experience that all things and events are self-expressions (*ātman*) of Brahman; we must not confuse with reality our ego-self (*jīva*) or our sense of "I" as distinguished from other selves or other things and events. Absolutizing one's ego-self in separation from other ego-selves constitutes a delusion (*māyā*) that makes one or one's community selfish and noncompassionate. In more Western terms, ontologically compassion arises from awareness that nothing is ever separate from anyone or anything else. For realizing that another human being's suffering is indeed our own engenders compassionate action to help that person or creature find release from suffering.

The Sanskrit word most often used for "compassion" is *karunā*, and all the ideals of Hindu wisdom regarding "compassion" can be summarized in one word: *ahimsā*, meaning "noninjury" or "nonviolence." Knowing by experience that all life is interdependent is the subjective source of a compassionate mind, the external expression of which is nonviolet interaction with all human beings and sentient life forms with which we share existence on planet Earth. Furthermore, compassionate nonviolence extends far beyond avoiding causing physical harm; it also includes not causing harm through speech and thought.

The best-known contemporary advocate of compassionate nonviolence is Mahatma Gandhi. According to Gandhi, compassionate nonviolence is noninjury in mind, speech, and action toward any creature. Specifically, this means (1) in mind, not to think maliciously of others; (2) in speech, not to use foul language—but more than this, to use language in a disciplined way as a vehicle to gently communicate ideas and values relative to the abilities and mental capacities of one's hearers; and (3) to avoid actions that do injury to any person or creature or community of persons and creatures.

THE BUDDHIST WAY

Like Islam and the Christian Way, the Buddhist Way traces its origins to a historical founder, Siddhartha Gautama, who 2500 years ago sat under a tree, meditated, and achieved an awakening experience that the earliest Buddhist texts describe as Nirvana, an "Awakening" to the interdependent structure of existence that transformed Gautama into the Buddha, the "Awakened One." Buddhists have sought to achieve a similar awakening experience by imitating the Buddha's example as portrayed in the plurality of Buddhist texts, schools, and meditative practices that together constitute the pluralism of the Buddhist Way.

One of the things we know with some certainty is that the historical Buddha taught that to realize Awakening the earnest seeker must develop two qualities: wisdom and compassion, often compared to two wings that work together enabling birds to fly, or two eyes that work together to see one's surroundings. By the standards of most Western philosophical and theological traditions, wisdom is understood to be something primarily practical and intellectual while compassion is something primarily emotional, and that wisdom and compassion are separate and incompatible. Since the beginning of the European Enlightenment, the natural sciences, presupposing Cartesian philosophical dualism, have generally concluded that the fuzzy, sappy emotion of compassion gets in the way of clear, logical, rational wisdom. This Cartesian dualism is still maintained in the ghettoizing of academic disciplines in contemporary Western universities into separate multiple disciplines that provide little, if any, integration of the research of these separate disciplines.

All schools of Buddhism, in their own ways, are theoretical interpretations of a core worldview, namely, that all existence is implicated in suffering and impermanence, that we cause our own suffering and the suffering of other sentient beings by clinging to permanent selfhood, that release from suffering is possible through ethical discipline based on the practice of compassion toward all sentient beings, and that the achievement of Awakening comes through the disciplines of meditation. These teachings, called the Four Noble Truths, are presupposed in every aspect of Buddhist teaching and practice even as they are nuanced differently in the various schools of the Buddhist Way.

The doctrine of nonself means that we are not embodiments of an unchanging self-entity remaining self-identical through time. Permanent selfhood is an illusion. What we "are" is a system of interdependent relationships—physical, psychological, historical, cultural, spiritual—that, in interdependence with everything else undergoing change in the universe, continuously creates "who" we are from moment to moment throughout our lifetimes. We are not permanent selves that *have* these interdependent relationships; we *are* these interdependent relationships as we consciously and unconsciously experience them. Because these relationships are not permanent, neither are we, nor anything else in the universe is permanent.

Furthermore, all schools of the Buddhist Way affirm that wisdom and compassion are "nondual," meaning utterly interdependent. The Sanskrit word translated as "wisdom" is *prājñā* (*panna* in Pali). The word translated as "compassion" is *karunā*. "Wisdom" is the achievement of Awakening, waking up to the nondual, interdependent, nonself structure of existence, which in turn a generates a compassionate mind responsive to the suffering of all sentient beings and a corresponding ability to work for the release from suffering of all sentient beings. To the degree that we have not awakened to nonself, to that degree our actions are selfish and compassionate action is confused with self-interest.

But perhaps the most widely known Buddhist tradition that grounds its teachings and practices on compassion is Japanese Pure Land tradition, particularly the "True Pure Land School" (Jōdō Shinshū) established by Shinran Shōnin (1163–1263). For Shinran, human nature is so corrupted by the fact of rebirth in the present age of *mappō* ("end of the Dharma") that no one can achieve Awakening by means of "self-powered" (*jiriki*) practices like meditation and devotion to the Buddhas and Bodhisattvas of Mahayana tradition. Humanity's only option is reliance on the "other-power" of Amida Buddha's compassion. Such reliance on Amida Buddha's other-power is not a self-powered act of trust but instead an unmerited gift transferred to faithful persons by means of Amida's compassionate "other-power" (*tariki*). Persons are thereby empowered to live compassionately in the world, free from the anxiety generated by self-powered attempts to achieve what one thinks one does not already have. At death, faithful persons are reborn into Amida Buddha's Land, called "the Pure Land," *as* Awakened Buddhas.[10] Thus Shinran's Buddhist Way can be summarized as "awakening by faith through Amida's other-power alone," where "other-power" and "grace"—in Luther's wording, "salvation by grace by faith alone"—are congruent notions.[11]

Of course Amida Buddha is not "God." And "sin" and *mappō* are not identical concepts, nor does the concept of *mappō* play the same role in contemporary Shin Buddhist tradition as it did for Shinran. Even so, the meaning of "faith" in both Shin Buddhism and Protestant tradition entails nontrust in one's self-efforts, which is not to say that Shinran was "the Luther of Japan." Japanese Pure Land Buddhism and Protestant, particularly Lutheran, doctrines of grace and faith may point to structurally similar experiences. But the objects of Pure Land faith and Christian faith are not identical. Nevertheless an experiential doorway exists where Buddhists and Christians can meet in interreligious dialogue.[12]

THE CHINESE WAY

The religious history of China is incredibly complex and not simply reducible to separate strains of Daoist and Confucian tradition. Rather it is more accurate to say that since the eighth century BCE the Chinese Way evolved from a blend of Daoist and Confucian strands into the Chinese Way, a Way of living in balanced harmony with nature within human community.[13] Accordingly, while the Daoist Way and the Confucian Way are utterly interdependent, it will be necessary to briefly treat each separately while keeping

10. Unno, *River of Fire*.

11. Ingram, *Dharma of Faith*, ch. 6.

12. Ingram, *Passing Over and Returning*, 67; Ingram, *Dharma of Faith*, ch. 4.

13. Ingram, *Passing Over and Returning*, 69–72.

in mind that the Chinese Way of seeking harmony flows back and forth between both traditions, seeking balanced harmony between each.

According to the *Dao De Ching* (Classic on the Way and Its Power), all living beings emanate from the *dao* (Way) and as such "incarnate" the *dao* in their own distinctive ways. So because the *dao* is structurally reflected in all human beings, there is always a possibility that it can emerge and express itself as virtue in the life of the individual so that all human beings—indeed, all sentient beings—are worthy of compassion.

The *Dao De Jing* devotes approximately half its chapters to the meaning of *dao* (chapters 1 through 37) and half to *de*, meaning "power" or "virtue," perhaps "the power of virtue" (chapters 38 through 81). Chapter 67 directly addresses compassion and its connections with frugality and humility. It reads:

> Every one under heaven says that our Way is greatly like folly. But it is just because it is great, that it seems like folly—well there can be no question about *their* smallness! Here are my three treasures. Guard and keep them! The first is "compassion"; the second, frugality; the third, the refusal to be "foremost of all things under heaven." Only he that is "compassionate" is truly able to be brave; only he that is frugal is truly able to be profuse. Only he who refuses to be foremost of all things is truly able to become the chief of all Ministers. At present your bravery is not based on "compassion," nor your profusion on frugality, nor your vanguard on your rear; and this is death. But "compassion" cannot fight without conquering or guard without saving. Heaven arms with "compassion" those whom it would not see destroyed.[14]

I interpret this text to mean that "compassion" is preeminent of the three virtues, and "frugality" and "humility" are its interdependent communal expressions. Compassion, frugality, and humility articulate the positive and negative consequences of acting with and without compassion: if you are compassionate, you will win, you will stand firm, and you will be protected. If you do not practice compassion, the consequence is death. Of course this can and has been interpreted in a number of ways, but I believe this is a reference to the Daoist view that sages through achieving union with the Dao by means of meditative and physical discipline achieve immortality. Since the Dao is ingredient in all things and events, knowing

---

14. Waley, *Way and Its Power*, 225. Waley used "pity" as his translation of the Chinese character *min*. I have translated *min* as "compassion," which I believe in a more accurate translation. See Welch, *Taoism*, 21–22; 43–44.

by experience rather than by mere intellectual assent to Daoist teaching empowers the sage's compassionate responses to whatever situation he or she encounters: the sage takes on qualities of a skillful teacher, who is able to communicate knowledge to students at their levels of experience so as to help students grow beyond their levels of experience. Only sages can be fully compassionate, because of their experiential union with the Dao. Nonsages can only imperfectly and approximately lead compassionate lives.

For Confucius, the Dao also referred to the balanced harmony that nature always seeks between opposite forces called *yin* and *yang*. But unlike the "philosophical Daoism" of the *Dao De Ching*, Confucius's focus was on balancing political and social forces necessary for the creation of compassionate community. His question, and the question of subsequent Confucian tradition, was, how can human beings learn to live peacefully in communal harmony with the Dao?

Confucius's Way centered on the concept of *ren*. The Chinese character for *ren* is written with the character for "man" and the number "two"—two men, or, less patriarchically, "two human beings together." Thus often translated as "human nature" and "humanity," *ren* also means "compassion." For Confucius, *ren* is the Dao specifically forming the "human nature" or "humanity" that identifies what it means to be a human being. Thus all human beings incarnate the Dao in their own distinctive ways. So the key to acting humanely and compassionately is knowledge by experience of our own nature as constituted by *ren*, which in turn allows us to know by experience the "humanity" of others and respond accordingly. For Confucius and subsequent Confucian tradition, compassion is an expression of the depth of our knowledge of *ren*, which transforms us into "Superior Human Beings" who are, as Mengzi (Mencius) put it, "sagely within and kingly without."[15]

Learning to become a compassionate human being is a lifelong process. Confucius's collective "sayings" preserved in the *Analects* are full of discourses on learning *wen*, or "the arts of peace," which is analogous to what Western universities and colleges refer to as the liberal arts. Art, music, literature, history, the study of nature, politics, economics are examples of activities unique to human beings. Studying and reflecting on the arts of peace reveal not only the uniqueness of human nature as constituted by *ren* but also our connective interdependence with all human beings. For Confucius, study and reflection are the only means by which to know by

---

15. Mengzi is regarded, as the "second founder" of Confucian tradition. See Lau, trans., *Mencius*.

experience what it means to be human and how we are interdependent with all human beings—indeed with all life forms as manifestations of the Dao. So the superior human being not only studies art but engages in artistic expression; the superior human being not only studies music but learns to play instruments, sing, and write music; the superior human being not only studies history but makes history; the superior human being not only studies literature but writes novels and poetry; the superior human being not only studies nature but applies knowledge of natural processes to creating technological structures that help to create human community; the superior human being not only studies politics and economics but engages in the rough-and-tumble of political and economic processes necessary for building communal life.

Accordingly, the Confucian Way is a Way of compassionate social engagement with the hard realities of politics and government required for creating communities of human beings living together peacefully and harmoniously. For study and application of the arts of peace reveal that all human beings are interdependent and interconnected—again, a notion very much in common with Western traditions of liberal arts education and contemporary natural sciences.[16]

CONCLUDING OBSERVATIONS

The religious Ways I have summarized in this chapter, plus those I have not mentioned—Native American and other aboriginal traditions plus Shintō, for example—all attempt to foster compassion as a means of linking human beings together in the creation of community and as a way of life that opens human beings to the presence of the Sacred "in, with, and under" all natural things and events, however this Reality is named by the various religious Ways. Of course, the meaning and purpose of compassion is tweaked differently in each tradition. Nevertheless, the religious Ways cited in this book are in their own plualistic ways unanimous in asserting the important role of the practice of compassion for the establishment of "the common good" as a means of creating human community—and human community with the life forms with which human beings must share planet Earth—and as a means of apprehending our connection with the Sacred.[17]

16. See Ingram, *Passing Over and Returning*, 81.

17. Cobb, *Sustaining the Common Good*, preface.

But compassion alone is powerless to resist unjust domination systems that have been the plague of human history. Contemporary examples abound: three hundred years of slavery in the United States; the oppression of Jews in Europe beginning in the fourteenth century and culminating in death camps set up by the Nazis and the resulting Holocaust; human sex trafficking; the Bush administration's war in Iraq, which was justified by the pretext that weapons of mass destruction were about to be launched on other countries in the Middle East and perhaps on the United States by Saddam Husain; the destruction of the earth's ecosystems and global warming caused by global corporations—mostly American and European—for the profit of the few at the expense of the poor and powerless. The earth calls for a justice that compassion alone seems unable to address. So the question is, what is the relation between compassion and justice in the religious Ways of humanity? Are compassion and justice dualities? Or are they interdependent, as *yin* is with *yang* in Daoist tradition? For compassion without justice is mere sentimentalism, while justice without compassion too easily degenerates into tyranny. What, then, do the religious Ways cited in this book tell us about the interdependence between compassion and justice? This question is the focus of the following chapter.

2

The Way of Justice

Compassion, the ability to "feel with" the other, is affirmed by the majority of humanity's religious Ways. Compassion is an ancient virtue that seems to have evolved among human beings at least since ancient hunters painted the shapes of animals on the walls of caves in France forty thousand years ago. Compassion seems to be part of humanity's DNA. Which does not mean that human beings have always acted compassionately. Compassion, more often than not is in continual struggle with another human evolutionary trait, "the survival of the fittest." And of course there are nuances that reflect the unique historical and cultural circumstances of each specific religious Way. Nevertheless there exists near universal agreement that compassion requires dethroning one's self from the center of one's world and putting "the other" in the center. For example, no one has described the universality of the obligation to practice compassion in Confucius's Way more clearly than Karen Armstrong:

> Compassion would become the central practice of the religious quest. One of the first people to make it crystal clear that holiness was inseparable from altruism was the Chinese sage Confucius (551–479). He preferred not to speak of the divine because it lay beyond the competence of language, and theological chatter was a distraction from the real business of religion . . . There were no abstruse metaphysics; everything always comes back to the importance of treating others with absolute respect . . . They should look into their own hearts, discover what gave them pain, and

then refuse under any circumstances whatsoever to inflict pain on anybody else.[1]

Three hundred years earlier the Israelite and Judahite prophets proclaimed the practice of compassionate justice as an absolute obligation between human beings and humanity's relationship with God. For the Israelite and Judahite prophets, compassion and justice were utterly interdependent to such a degree that neither could exist apart from the other. While the meaning of "justice" will be the focus of this chapter, it must be kept in mind that compassion and justice are two sides of one coin.

THE JEWISH WAY

The eighth-, seventh-, and sixth-century Israelite and Judahite prophets who proclaimed the obligation to live justly and compassionately in community reflect a long "process of monotheizing" that evolved from the stories of the exodus of Israelite slaves from Egypt under the leadership of Moses and their meeting with God on Mount Sinai as recorded in Exodus 1–20. This process further evolved in the prophetic oracles recorded in Jeremiah and Ezekiel, where Yahweh is explicitly proclaimed as the only deity that exists. The monotheizing process was continued by the rabbinic tradition, particularly after the disappearance of the temple priesthood with the fall of Jerusalem in 70 CE, and by the early Jesus movement.[2] Accordingly, relating to all human beings compassionately and justly is a central commandment of the Torah (*Torah* here means God's "instructions," not "law"). The obligation to live compassionately and justly is derived from the only God there is. As the Shema declares: "Hear O Israel, the LORD our God, the LORD is One" (Deut 6:4).

This continuing consciousness of God's covenant with Israel also means that Jews are forever obligated take care of one another as well as persons who are not Jewish—not only when Jews lived in close proximity to one another, but also when they became aware of Jews in distress in other locations. During the time that Jews lacked political sovereignty, they became a community grounded in a shared historical memory of the Exodus narratives and a shared communal history. They believed that the fate of the Jewish people, regardless of temporal domicile, was linked. This is one

---

1. Armstrong, *Case for God*, 22–23.
2. See Sanders, *Monotheizing Process*, ch. 1

of many convictions that help explain the success of the Zionist movement, the historically unprecedented resurrection of national identity and political sovereignty after two thousand years of dispersion. The monotheistic consciousness that evolved out of the Exodus traditions as continued by the prophets was the glue that held the Jewish people together even through such disasters as the Holocaust. It was the foundation of Jewish survival as the most persecuted religious community in the history of the world's religious Ways.

The Hebrew word for "justice," as I noted in the previous chapter, is *mišpaṭ*. The best literal translation of *mišpaṭ* is "right treatment of people," which means giving human beings what is needed for meaningful existence, which is often not what human beings want. Justice, *mišpaṭ*, is always interdependent with *ḥesed*—"compassion": the experience of someone else's or another community's suffering, injustice, poverty, or exploitation as one's own, and doing justice on behalf of that person or community. But there is an ambiguity in the polar relationship between justice and compassion. Often, too often, standing up for justice has been taken to imply directly resisting (often violently) unjust persons or unjust political, social, or religious domination systems in which a small minority of wealthy and powerful people exploit a large majority of human beings for their own political and economic benefit. In such cases, justice can seem divorced from compassion. Violent action to defeat the unjust often places compassion on hold. In such cases, compassion linked to justice is demanded for those whom justice requires defending. But justice not linked with compassion is too often a necessity for resisting and defeating injustice.

No religious Way has experienced this ambiguity more profoundly than the Jewish Way because no religious community has suffered such harsh persecution. The Torah demands following God's commandment to live compassionately and justly. But within the ever-changing historical and cultural contexts of Jewish history, the perennial question is, how? This question has been the center of rabbinic reflection going back to the foundational narratives recorded in the Torah—that is, the first five books of the Tanak.

THE CHRISTIAN WAY

The historical Jesus was a Galilean-Judean mystic and teacher of subversive wisdom that called for resistance against the exploitative political,

economic, and religious domination systems that oppressed the vast majority of human beings in his time and place.[3] Like the eighth-, seventh-, and
sixth-century-BCE prophets who preceded him, Jesus spoke of *mišpaṭ* and
ḥesed. *Mišpaṭ* occurs more than 200 times in the Tanak, where "justice"
means acquitting or punishing persons on the merits of the case, regardless of social standing or religious identity. But for Jesus, again in harmony
with the Exodus and prophetic traditions, justice meant more than merely
forensically giving human beings their rights (that is, giving people their
due, whether punishment or protection or care).

The historical Jesus also stressed God's "compassion" (*ḥesed*) mediated
toward those whose deeds merited punishment according to conventional
rules imposed by powerful religious authorities (the temple priests and the
legal scholars referred to as "scribes") and political authorities (the Romans
and the Judeans and Galileans who worked for Rome) upon the vast majority of the poor these authorities oppressed for their own economic and
political benefit. A more simple way of stating this is that Jesus explicitly
taught that God's mercy and justice are preferentially directed toward the
poor and the marginalized, while God's justice is preferentially directed
towards conventional politicians and the wealthy oppressing the poor.[4]

Saint Paul, who transformed the Jesus movement into a Christian
Way as it began to separate from the Jewish Way, was himself a diaspora
Jew whose original name was Saul of Tarsus. He was an avid persecutor
of the Jesus movement shortly after Jesus's crucifixion around 30 CE, but
because of a conversion experience (Acts 9:1–18), he undertook a mission
to spread his interpretation of the Jesus Way to Gentiles. Paul's mission to
the Gentiles was thus the beginning of a gradual separation between the
first-century Jewish Way and an emerging Christian Way. It is in this sense
that Paul is the "founder" of the Christian Way.

But the separation between the Christian and Jewish Ways was not
complete until the fourth century, when Emperor Constantine declared the
Christian movement to be the official religious Way of the Roman Empire.
By this time there were several competing versions of the Christian Way,
so that Constantine felt compelled to call the Council of Nicaea into session to hammer out which form of the Christian Way would be the official
religious Way of his empire. Christian theologians have been debating how

3. See Borg, *Jesus*; Oakman, *Political Aims of Jesus*, ch. 5.

4. For example, see See Matt 5:3–12; Matt 5:38–42 = Luke 6:29–30; Matt 5:43–48; =
Luke 6:27, 32–36; Matt 6:19–29 = Luke 12:33–34; Matt 6:25 = Luke 12:21–31.

to authenticate an official or "orthodox" version of the Christian Way, particularly the relationship of the historical Jesus to God, ever since. After the Council of Nicaea Saint Paul's teaching of "justification by grace through faith" was again affirmed by Saint Augustine and later by Martin Luther, who added the word "alone" to this formulation. Paul's and Augustine's notions of justification has dominated mainline Protestant movements since the sixteenth century, and while prevalent in Roman Catholic and Orthodox traditions, this teaching has received much less emphasis.

In its original context in Paul and Augustine's teachings, "justification" is a legal term originating in Roman law and names the process by which a Roman judge in a court of law could declare a criminal legally guilty of a crime, but not impose a penalty. In Paul's and Augustine's theological interpretation of this legal concept, justification signified the process by which through grace God forgives persons for sin, for which God's justice requires the penalty of death, so that sinful human beings are thereafter free—not from sin—but enabled to compassionately treat human beings in a way similar to the way God has treated them. But God's justice must first be satisfied. The satisfaction of God's justice is accomplished through the death and resurrection of the historical Jesus confessed to be the Christ of faith, which allows God to be compassionate to sinful human beings.

This doctrine has caused much debate in the two-thousand-year history of Christian theological reflection. For example, traditional Roman Catholic theology and practice tweaks this teaching much differently than mainline Protestant interpretations; contemporary feminists reject this aspect of Paul's and Augustine's theology as "theology of divine child abuse," a position with which I agree. And there is no universal agreement regarding exactly how Paul's or Augustine's teachings should be applied throughout the changing epochs of Christian history. But this much seems clear, at least to me: compassion wherever possible and justice wherever possible go beyond legalism and are the nondual foundations of authentic Christian faith and practice. This is so because God, as portrayed in contemporary Christian liberation theology, is best characterized by compassionate and just preference for the poor and oppressed.

THE ISLAMIC WAY

The relation between justice and compassion is a major theme in the Qur'an, the Sunna (the "Custom of the Prophet Mohammed"), and the legal traditions recorded in the Shari'a (something like "the foundations of law") upon which legal decisions (*fiqh*) are made, according to which Muslim communities and individuals are to test and measure their surrender to God's will within the ever-shifting demands of fourteen hundred years of history. Compassion and justice are two of the "ninety-nine beautiful names" of Allah recorded in the Qur'an.

There are two interdependent ways justice is understood in the Qur'an and Islam's legal traditions, collectively referred to as "Shari'a." As a legal principle, the Qur'an specifies not only how Muslims are to conduct their lives, but also how they are to structure relationships with other Muslims and non-Muslims. The Qur'an also specifies punishments for specific crimes, coupled with justification for these punishments. Furthermore, the Qur'an and the Sunna specify that human beings who act justly (and compassionately) will be rewarded with paradise after their deaths.

But within the House of Islam there is a long history of debate about exactly how justice can be fulfilled within the ever-shifting cultural and historical contexts of Islamic history. What might appear just in Mohammed's day and time might not be so today. So while all Muslims agree that faithful Muslims should live justly, the question facing Muslims for fourteen hundred years is, how? While Muslims agree with the Qur'an's declaration that God will never do injustice, how justice is expressed between Muslims and between Muslims non-Muslims is a continual topic of conversation.

Regarding divine justice Muslims throughout history have debated the relations of divine justice to justice between human beings and how divine justice will be fulfilled in the future, both within and outside the House of Islam, even as there is agreement that God will in no case act unjustly. There is also a long history of debate among the Ulama (scholars legal, theological, and philosophical) about justice for non-Muslims. Although the Qur'an is not all that specific about God's justice for non-Muslims, in three instances it clearly states that the good deeds of persons belonging to non-Muslim communities, particularly Jewish and Christian communities, are rewarded by God.[5]

As I previously noted, in the Qur'an's original language, "justice" was a general term that applied to individuals and reflected the nomadic, tribal culture in which Mohammed lived and taught. But over time, Islamic jurists sought to unify political, legal, and social justice as important themes in Islamic jurisprudence. In general, the schools of Islamic law stressed the

---

5. For example, see Surahs 1:147; 4:58; 17:20–21.

exercise of reason, free will, and responsibility. Two concepts guide Islamic legal reflection on justice and its application: (1) *huquq*, meaning "rights," and (2) *hsan*, meaning "generosity" beyond obligation. These two concepts set the standard guidelines by which Muslims individually and collectively seek to structure individual and social life on the foundations of justice.

Accordingly, the Qur'an's demand that persons govern their lives by justice transcends all social boundaries and is required of all human beings, whether they are Muslims or non-Muslims. The Qur'an also states that one's condition in the afterlife will be measured by God to the degree that one has affirmed the unity, compassion, and justice of God for all creation and has acted accordingly toward other human beings.[6]

THE HINDU WAY

The plurality of Hindu scriptural traditions reflects a deep fascination with the meaning of justice, both as a social reality and a cosmic principle. The earliest Vedic narratives identify justice with the work of a deity such as Yama, who weighs the actions of the dead on his scale, or Varuna, who punishes unjust human beings with illness. By the end of the Vedic period (sixth century BCE), justice was equated with a cosmological principle called *rita*, which governed nature as well as human ethical conduct. To follow *rita* was to act in accordance with justice.

But it was not until the concept of karma emerged in the early Upanishads that justice was understood as a logical consequence of human action. Karma stipulated that good actions are rewarded and bad actions punished, if not in this life then in one's next rebirth. In later centuries, "justice" was also one of the many meanings as *dharma* and played a major role in the social and political order. Ideally embodied in the person of the king, justice became a leveling tool and a means of protecting the weak from the strong. And although Hindu society was divided into ranked castes with distinct duties and rights, a universal respect for foundational values such as justice pervaded the entire social structure.

Similar to Platonic notions of justice, the Hindu Way has since ancient times linked justice with the performance of duties prescribed by *dharma*. The Hindu social and political activist Mahatma Gandhi (1869–1948) placed the ancient traditions of justice at the center of his social movement

---

6. See Surah 4:135; 35:28; 49:9; 58:11.

that simultaneously forced the British to leave India and reflected his vision of how India could be transformed into a modern democratic society.

Mohandas K. Gandhi, most widely know as "Mahatma" or "Great Soul," did not explicitly address the concept of justice philosophically or theologically and there are no discussions in his writings on theoretical concepts of social justice. But justice as rooted in the scriptural and philosophical schools of the Hindu Way in dialogue with Western notions of justice provided the foundations of his social activism against unjust political and economic systems. As one trained in the British legal system, Gandhi was the pioneer of the movement for social justice in India. Of course, before Gandhi, Hindu poets, Jain saints, and other social reformers paid close attention to the social injustice issues originating in caste discriminations and the practice of untouchability. Ancient Hindu and Jain notions of "truth-force" (*satyāgraha*) and nonviolence (*ahimsā*) were the guiding concepts of Gandhi's social activism to free India from British rule, of his attempts to create a state that would meaningfully include all religious Ways, including Islam, and his continued support of Muslims after the creation of the state of Pakistan.

Gandhi was born into the Vaiśya caste (merchants, peasants, what might be roughly called "working class") in a Vaisnavite family (devotees of Vishnu) with Jain friends, who greatly influenced his application of nonviolence to the struggle for social justice in a state free from oppressive British rule. Fresh out of law school in England, he went to South Africa in 1891, where he defended Muslims in a court case, and where his experience of and resistance to racial discrimination and oppressive government began. After his return to India he founded the Natal Indian Congress, where he conjoined nonviolence and "truth-force" with *tapaysa* (renunciation) and *swarj* ("self-rule) as guiding principles of the Indian Independence Movement.

But Gandhi's insistence on interreligious dialogue and inclusiveness created suspicion among conservative Hindus (the Araya Samaj, or "Society of Noble Ones"), who desired to establish their version of Hindu orthodoxy as the state religion of an independent India. While Gandhi was on his way a prayer meeting celebrating India's independence in 1948, a member of this movement assassinated him. He died uttering the name of Rama, one of the many *avatars* ("descents" or "incarnations" of Brahman) according to one of Gandhi's favorite texts, the *Bhagavad-gītā* (Song of the Blessed One). Following Gandhi's assassination, Vinoba Bhave continued his work, particularly in transforming Indian village life. It should be noted that Gandhi and Bhave worked to outlaw the caste system that according to their view was an oppressive social system as unjust as British colonialism.

THE BUDDHIST WAY

Classical Buddhist approaches to justice begin with individual behavior and center on the law of karma, in which good actions generate positive consequences and bad actions negative consequences. Consequently, the Buddhist Way has proved historically compatible with any number of different political systems. Because the Buddhist Way has traditionally emphasized monastic life and discipline, Buddhist doctrines and practices have focused on general social prescriptions—the five precepts of good conduct (not to kill, steal, lie, commit immoral sexual acts, or partake of intoxicants)—while acknowledging the existing political systems of the cultures to which the Buddhist Way has been transmitted. Rulers, in turn, have often patronized the *sangha*,[7] providing a mixture of protection and resources in return for the blessings of the monks and the wider political legitimacy it afforded them. These basic arrangements originated with King Ashoka on the Indian subcontinent in the third century BCE and continue through many contemporary democratic and autocratic regimes in Buddhist-majority countries. The last two decades have seen the growth of socially engaged Buddhism in the United States, Europe, and Southeast Asia (Vietnam, Cambodia, and Burma), along with support for the Dalai Lama and for greater Tibetan autonomy from China among Buddhists living in North America and Europe.

The ultimate goal of all Buddhist practice is the attainment of Awakening, often described as "liberation" or *nirvanā*, and often interpreted as a state of being "extinguished," as in the extinguishing of a fire. Thus *nirvanā* refers to the elimination or extinguishing of various mental obstacles or "defilements" that block an individual's attainment of Awakening—obstacles derived from the three "poisons" of desire, hatred, and ignorance. In general Mahayana Buddhist teachings, Awakening has a slightly broader

---

7. In the Theravada Buddhism rooted in South Asia, the monastic community; in the Mahayana Buddhist Way originating in East Asia, the entire community of Buddhists, although primary emphasis is still given to communities of monks and nuns living the monastic life.

meaning, referring to the attainment of "wisdom" (*prājñā*), which is interpreted as freedom from the recycling bondage of life and death.

Buddhist emphasis on individual Awakening is often portrayed as quietist. But historically, this is not an accurate characterization. Nevertheless, it is true that even today Buddhist thinkers rarely address social justice issues such as human rights, the fair distribution of resources, the impartial rule of law, and political freedom. Of course, the Buddhist Way is hardly alone in this regard. Almost all the ancient philosophies and religions paid scant attention to issues of social justice. Even Catholicism, which has addressed social issues from early times, did not concern itself with questions of social justice or use this term in official documents until the latter part of the nineteenth century. Indeed, not until the eighteenth century did social justice emerged as an important issue in political thought and social philosophy in the West. The last three centuries have thus seen the maturation of such key concepts as citizenship, political equality, and the fair distribution of economic resources. However, the process of modernization that drove the development of social philosophy in the West paradoxically retarded it in the East. Belatedly experiencing modernization as an expression of Western imperialism initiated by military and economic contact with Western colonial powers, Eastern intellectuals lost confidence in their native traditions, coming to see them as relics of the past without relevance to contemporary problems. As a result, indigenous philosophies and religions tended to be neglected in favor of the study of Western thought.

This process has only recently begun to reverse itself. As South and East Asian Buddhists became increasingly aware of the value of their particular cultural and religious identities, a new strain of Buddhist thought began to emerge, interested not only in relating the Buddhist Way to modern concerns, but also in exploring the applicability of Buddhist notions of nonviolence to contemporary justice issues. Without doubt, the most important of these Buddhist groups is Socially Engaged Buddhism.

The term "Socially Engaged Dialogue" was first used as a description of Buddhist traditions of social activism by Sallie B. King in her analysis of Thích Nhất Hạnh's notion that "inner work," or meditation, must engender nonviolent "outer work," which King describes as "social engagement" with the systemic structures of injustice.[8] Buddhist–Christian conceptual dialogue has generated deep interest in the relevance of dialogue for is-

---

8. King, "Thích Nhất Hạnh and the Unified Buddhist Church"; and King, "Conclusion." Also see Nhất Hạnh, *Interbeing*.

sues of social, environmental, economic, and gender justice. Because these issues are systemic, global, interconnected, and interdependent, they are neither religion specific nor culture specific. Accordingly, socially engaged dialogue is concerned with how Buddhists and Christians have mutually apprehended common experiences and resources for working together to help human beings liberate themselves and nature from the global forces of systemic oppression.

But dialogue with Christian liberation theology has raised questions about the central role the practice of nonviolence has played in Buddhism for twenty-five hundred years and whether Buddhists can develop their own distinctive understanding of justice. The key point is liberation theology's strong insistence on political, social, economic, and environmental justice, whereas traditional Buddhist teaching seems to have little to say about justice, in the sense that these teachings do not normally employ justice language. But there are well-known exceptions like B. R. Ambedkar and Sulak Sivaraksa, who never shy away from justice language.

Of course the question is, how exactly should the practice of nonviolence, which all religious Ways encourage, be related to the equally important requirement for justice in human and communal relationships? Lamenting that Buddhist emphasis on the practice of nonviolence has little, if any, relevance to violent injustice done to Jews in Nazi Germany or to Cambodians murdered in the "killing fields of that country, King, who is a practicing Zen Buddhist and Quaker, seeks a Buddhist "middle way" of joining nonviolence to the practice of justice. In a lecture given at the 2014 Annual Meeting of the Society for Buddhist–Christian Studies, she explains:

> Looking at both the Buddhist and Christian ways of thinking, it seems that what we are dealing with here is a virtue with the characteristic of an Aristotelian mean. Let us call this virtue, "critical voice." An excessive *amount* of "critical voice" is seen in strident, frequently angry and self-righteous use of the prophetic voice and an inability to hear the voice of the other side, even when there may be some truth on that side or a necessity to hear that side in order to resolve a situation. A *deficiency* of critical voice is seen in a failure to take on the prophetic voice when it is needed, a failure to criticize or condemn what should be condemned. It seems that Engaged Buddhism tilts to the one side (the deficiency in critical voice) and needs to hear the correction from Liberation Theology and some Liberation Theology needs to tilt to the other side (the

side of excessive critical voice) and needs to hear the correction from Engaged Buddhism.[9]

So here is the justice issue facing the Buddhist Way—a question, be it noted, not unique to the Buddhist Way. There has always existed in human history injustices so terrible, so destructive, so inhumane, so senseless, so rooted in human nature that nonviolent compassion seems an irrelevant method of confrontation. Examples abound throughout human history, but the Nazi Holocaust and the Killing Fields of Cambodia are contemporary examples. Confronting such collective evil requires the application of justice both to the perpetrators and those who have survived collective violent oppression. Hopefully, justice can be administered nonviolently and compassionately, but more often than not justice is violent in its application, as described in Christian just war theory. The question for all religious Ways is, exactly how should justice and compassion be balanced in confrontation with real issues of political, economic, and environmental injustice? For Buddhists, this question is made acute because there exists little interest in the concept of justice in the Buddhist world, with the possible exception of socially engaged Buddhists.

THE CHINESE WAY

One of the Daoist Way's most important concepts is *wu-wei*, which can be translated as "nonaction," nondoing," or perhaps "action without the clinging to the fruits of action." The classical symbol for *wu-wei* is water—not water, for example, in a swamp (although this form of water is certainly creative) but free-flowing water. As chapter 123 of the *Dao De Ching* (Classic on the Way and Its Power) describes it:

> Nothing under heaven is softer or more yielding than water; but when it attacks hard and resistant things there is not one of them that prevail. For they can find no way of altering it. That the yielding conquers the resistant and soft conquers the hard is a fact known to human beings, yet utilized by none.[10]

So "actionless action" is essentially the nonassertion of ego in action. That is, classical Daoist tradition sought to train people through systems of meditation and physical disciplines to let go of all ego-centered desires

9. King, "Through the Eyes of Auchwitz and the Killing Fields."
10. Waley, *The Way and Its Power*, 238.

in order to be naturally formed by the balancing processes of nature. Thus, as we gradually do nothing, the Dao gradually does everything through us. In this rather long process, we gradually acquire *de* or power, or more accurately, "the power of virtue."

Wu-wei is also the source of justice, according to the philosophical Daoist tradition. This is so because the Daoist Way arose as an answer to the social and political anarchy of the Warring States Period (475–221 BCE). This was a time of political and social turbulence. The dominant issue of these dangerous times was how human beings could learn to live peacefully together in community based on justice and harmony that mirrors the balancing flows of nature's processes. The classical Daoist Way asserted that all governmental systems are unnatural because they are artificial systems that force human to live in accordance with the collective egos of powerful political elites imposing their standards on the majority of the community. It is governments that start wars, create poverty, and unjustly oppress human beings by forcing them to act unnaturally, (i.e., out of harmony with the Dao). So "justice" means that rulers and politicians should govern by nongoverning in order that what is natural for human community can gradually evolve in harmony with the Dao.

The Confucian Way also based its distinctive teachings and practices on living in harmonious balance with the Dao, but in a way that sought to address the hard realities of politics and statecraft. Daoist tradition was predominately a way of withdrawal from social engagement as its practitioners left settled communities to live solitary monastic lives mostly in isolated, small mountain communities. Confucian tradition was and remains a way of social engagement. Since both traditions originated during the end of the Period of the Warring States, both sought to answer the question, How should human beings learn to live together in harmonious balance with the Dao"? They answered this question differently. But the Chinese people did not see the Confucian Way and the Daoist Way as dual opposites that required choosing one while rejecting the opposite. For if one thinks of the Daoist Way as a *yin* philosophy and the Confucian Way as a *yang* philosophy, the Chinese people sought to live in balanced harmony between them. There are times when one should follow the Confucian way primarily, and there are times it is wise to follow the Daoist Way. So the Chinese Way of religious faith and practice is a wise balancing act between the Daoist Way and the Confucian Way. So what are the distinctive Confucian contributions to the Chinese Way?

From its beginnings, the Confucian Way has addressed the question of just social and political order. Confucius (ca. 550–480 BCE) and his most influential followers, including Mengzi (ca. 370–290 BCE), attached great importance to proper and just relations between individuals and within families, communities, society, and the state. Historically, an emphasis on filial piety and reciprocity legitimated a patriarchal order in which the emperor was viewed as the father of his subjects, and the father as the master of his wife and family. Beginning with the Han Dynasty (206 BCE—220 CE), state-supported Confucian rites and sacrifices served to dramatize the harmony between heaven, earth, and humanity and underscored the justice (*yi*) of the emperor's rule. In Confucian teaching, justice is no abstraction but rather the concrete realization of harmony and reciprocity in communal relationships, a concept similar to process theology's idea of community based on "the common good."

Specifically, early Confucian notions of justice appropriated the ancient idea of "the Mandate of Heaven," whereby successful rulers who presided over peace and prosperity were seen to possess divine favor. Unjust rulers who did not fulfill their obligations to society were liable to fall upon personal and political misfortune—evidence of loss of Heaven's Mandate—that often lead to rebellion against such a ruler.

Obviously, the Confucian Way is not the mystical Way of the Daoists, but a Way of social engagement with the political and ethical necessities required for constructing human community. I am not the first one to notice this because, as noted above, the Chinese people viewed and still view both traditions through the filter of China's classical worldview. As a Way of inner experience and withdrawal from the issues of statecraft in order to follow the natural ebbs and flows of nature, the Chinese typically viewed the Daoist Way as a *yin* way of life. But as a Way of social engagement, the Confucian Way was regarded as *yang* Way of life. So, again as previously noted, the Chinese Way seeks to live at the harmonious balancing point between the Daoist Way and the Confucian Way. For the Chinese Way, justice only occurs by living in balanced harmony that includes elements of both the Confucian Way and the Daoist Way.

CONCLUDING OBSERVATIONS

While compassion lies at the heart of each of the religious Ways I have described in summary fashion, justice seems nuanced quite differently in each religious Way. Justice is a cornerstone of Jewish, Christian, Islamic, Hindu, and Confucian teaching and practice, but seems not to have been of much concern in the Buddhist Way until the beginning of the Buddhist–Christian dialogue that began at the University of Hawaii in 1980 and continues today in the Society for Buddhist–Christian Studies. Furthermore, the meaning of "justice" and how to achieve it varies from religious Way to religions Way. Engaged Buddhists seem very interested in Christian and Jewish traditions in this regard while Buddhist dialogue with Islam is also an increasing dialogical interest among socially engaged Buddhists for the same reason.

But there's a hiccup. "Compassion" and "justice" are meaningless abstractions unless in some way these concepts become living realities in the lives of human beings living in community. So the questions are what is the meaning of community, and how should compassion and justice be structured to create community based on what process theologians call "the common good"? The meaning of "community" and its relation to compassion and justice according to the religious Ways summarized in this book is the topic of the next chapter. The structure of compassionate and just community seeking the common good for all human beings, indeed for the life of planet Earth, will be focus of the remaining chapters.

3

The Way of Community

Webster's Third New International Dictionary of the English Language defines "community" as:

- A group of people who live in the same area, for example in a city, town, neighborhood, or village

- A group of people who share the same interests

- A nation or group of nations

- The whole human community

- The environing world

- A future reality to be achieved, for example a "Kingdom of God" in classical Christian theology[1]

What is striking about this list is the utter lack of attention to the interrelationship between compassion and justice in the formation of "community." According to the religious Ways surveyed in this book, without compassion and justice there can be no community. So what *Webster's Dictionary* has described, with the exception of the last item, is nothing more than groups of individuals having no responsibility for the welfare of other human beings who happen to be living in close proximity with one another.

Apart from concrete embodiment in human togetherness, compassion and justice are empty abstractions, at best unrealized possibilities within the rough-and-tumble of concrete existence characterized by Alfred

1. Gove, *Webster's Third New International Dictionary.*

Lord Tennyson's poem *In Memoriam* (1850) as "nature red in tooth in claw." Furthermore, "community" apart from compassion and justice is an equally meaningless abstraction. While the religious Ways that are the foci of this book have affirmed the interdependence of compassionate justice as essential for the building of community in ways particular to their worldviews, I have always thought that the Israelite and Judahite prophets spelled out with particular clarity how justice and compassion are the creative underpinnings of human togetherness-in-community. Or as the prophet Micah declared, we "have been told what is good" at all times and in all places.

But here's the hiccup: exactly *how* and *by what means* are the ideals of compassionate justice to be balanced with the hard political and economic demands of human togetherness? Exactly *how* should a mere "gathering of individual human beings" be transformed into a community constituted by compassionate justice for "the common good"?[2] Human beings have worked to live compassionately and justly in community throughout history, but the historical evidence indicates the utter failure to establish lasting human community anywhere on planet Earth. Like nature described by Lord Tennyson, human history is also "red tooth and claw."

Still, there have been attempts, mostly in humanity's religious Ways, to bring compassion and justice to bear on the creation of community. For this reason it might be well to listen to the collective voices of these Ways.

THE JEWISH WAY

One of the amazing facts that has emerged in the history of religions is the sheer existence of the Jewish Way, a three-thousand-year history of struggle to figure out the implications of the Torah for establishing a community founded on compassionate justice so as to be "a light to the nations" (Isa 49:6), a model upon which Gentiles might draw in creating compassionate, just communities relative to their own cultural contexts. The Jewish Way has been persecuted throughout its history for attempting to establish communities grounded in compassionate justice: by the Egyptians in Moses's time, by the Philistines and Assyrians in the eighth century BCE, by the Babylonians in the sixth century BCE, by the Greeks and the Romans, and particularly by Christians after Constantine (272–337) established the Christian Way as the state religion of the Roman Empire. From this time until the Second World War the main persecutors of Jews have been

2. See Cobb, *Sustaining the Common Good.*

Christians, by numerous methods so brutal that it is amazing that the Jewish Way has survived in such rich diversity as a "light to the nations." Persecution of Jews, anti-Semitism, is still practiced by conservative and fundamentalist Christians and by contemporary radical Islamist groups who ignore the Qur'an's injunction against persecuting human beings on the basis of religion.

To be a Jew means, among other things, continually wrestling with God's command that human beings must live in communities structured by compassionate justice. While the meanings of "compassion" and "justice" are quite clear in the prophetic traditions, the exact meaning of "community" and how to establish it has been an ongoing debate in the rabbinic opinions recorded in the Talmud. The debate continues in the twenty-first century. Community is structured differently in tribal societies than in agricultural and urban societies; "community" has different meanings in American or western European cultures than in Middle Eastern, South Asian, or Eastern cultures. In other words, "community" in Jewish experience names a plurality of structures of togetherness ranging from exclusivist to inclusivist to pluralist—all seeking grounding in compassion and justice. But compassion for whom? Justice for whom?

So why has the Jewish Way survived against all possible odds for over two thousand years? The answer lies in community as the central focus of the Jewish Way and in the prophetic demand for fair treatment of all members of that community, including workers, the poor, and resident aliens. The particular structures of Jewish community have changed throughout the centuries, even as God's commandment to established compassionately just communities has not changed.

The Jewish Way specifies that the ultimate purpose of community is to "repair the world." Jews engage in world repair by means of daily and weekly prayers focused on community. These prayers include the "Prayer of the Congregation" or tifilah l'kehilah, which originated among exiled Jewish communities resettled in Babylonia by Nebuchadnezzar in 597 BCE and again in 586 BCE. Living in the world's first ghettos, surrounded by a foreign, conquering culture, the Judahite people developed these prayers as means of maintaining their identity so as to be able to "worship God in a foreign land," as Ps 137:4 has it. The themes of the Prayers of the Congregation are the labor required to make community function in a compassionate and just way.

The second form of prayer for building community is called *aleinu*, meaning "our duty." *Aleinu* prayers present visions of the world as the world should be, which means as God intended the world to be at the time of creation but that now human beings have broken. *Aleinu* embodies the movement for *tikun olam*, human actions directed toward the repair of God's world.

There is a reality therapy at work in these prayers. The work of "repairing the world" is ongoing and has never been fully achieved in the long history of the Jewish Way even within specific Jewish communities. Yet the Jewish people keep trying, whether living in ghettoes in Babylonia or in medieval Europe or in every contemporary city or village where Jews live and work and die. There exists a wide body of rabbinic opinion that human action alone will not be enough, but that sometime in a future known only to God a worldwide community of compassionate justice will be finally fulfilled with the appearance of the Messiah, who will be a sign that a new age has begun when all human beings will live together as one for the common good. But until this event, human beings must struggle on, must keep on keeping on, in the effort to build community no matter how imperfect.

THE CHRISTIAN WAY

Koinonia is a Greek word that can be translated into English as "communion," "association," "fellowship," "sharing," "common," "contribution," and "partnership." Not one of these words, however, adequately captures what the early Christians meant when they spoke of the *koinonia* they had with one another and the historical Jesus as the Christ of faith. For the first generations of Christians, *koinonia* expressed a relationship of great intimacy and depth. The implications of this word are inclusive and describe the communal relationship of faithful Christians with God and other Christians and non-Christians.

The earliest model for a distinctively Christian *koinonia* is found in Mark 9:38:

> John said to him, "Teacher, we saw a man casting out demons in your name, an we forbade him, because he was not following us." But Jesus said, "Do not forbid him; for no one who does a mighty work in my name will be able to speak evil of me. For he who is not against us is for us."

The context of these verses is an account of when Jesus and his disciples returned to his home in Capernaum after an extended journey. So once more Jesus instructs the disciples about the meaning of community. His teaching method relied on subverting their notions of community as grounded in preeminence of position: "He who would be first must be last," he instructed in previous verses. But now the disciple John changes the question again by latching onto the name of Jesus as he brings up the issue of just who can be included in the Jesus community.

Nowhere in the Gospel of Mark—or in the other Gospels—does the historical Jesus recommend party identity as the standard of inclusion in the Jesus community; rather he opens up community to anyone who is not against it. There are no fixed criteria for membership in the Jesus community—beyond the requirement that one not be against it. There is no imagined criteria for inclusion in Jesus's *koinonia*, no gold-card standard of membership. The historical Jesus's teachings as recorded in Mark are pluralistic, not exclusivist; they apply to all who are not against Jesus—not only his disciples and followers but to Christians and non-Christians: Buddhists who revere the historical Jesus as a bodhisattva; Muslims who revere the historical Jesus as one of the greatest prophets; Jews who interpret the historical Jesus as reformer calling the people of his time and place to a renewed practice of Torah.

Of course, these non-Christians do not accept Christian *ideas* about the historical Jesus as the Christ of faith. But Jesus's teachings recorded in these verses in Mark make no such stipulation. To be *for* Jesus does not necessarily mean accepting ideas *about* Jesus—Christian creeds, doctrines, theological constructions in general—all of which flow out of conventional wisdom and are tied to historical and cultural contexts and are thereby empty of unchanging essence and once-and-for-all timeless meanings. The only obligations for membership are feeding the poor, helping the homeless, healing the sick, and engaging in struggle against the political, economic, and religious domination systems that keep human beings in poverty, poor health, and hunger. A member of the Jesus community, at lest in Mark's Gospel, is not one who belongs to "proper" groups. Anyone who is not against the "historical Jesus" is a follower of the historical Jesus's community, which makes for a very pluralistic community indeed.

Likewise throughout Saint Paul's epistles, "community" is all-inclusive, deep, personal, and intimate and entails sharing one common life within "the body of Christ" at all levels of experience and existence—spiritual, social, intellectual, economic, and political. All areas of life are to be included. A Christian community is not simply a society or a fraternity. It is an *ekklesia* or "church" that exists *in* God and *with* God established *by* God's grace as incarnated in history in the life, death, and resurrection of the historical Jesus confessed to be the "Christ" ("messiah") of faith.[3]

So from the very beginning of its history as a subsect within Judaism through the gradual emergence of distinctive Christian communities separating from the Jewish Ways the early church tried to be a community seeking justice and compassion for all persons as it struggled to understand and define what "community" as "the body of Christ" meant. The Christian Way from its beginnings was a corporate community; to be "Christian" meant to belong to the *koinonia*. No one can be "Christian" as an isolated individual, but only together with, as Saint Paul phrased it, "the brethren" in a "togetherness" of a "common life" within "the body of Christ."[4] Even the early Christian hermits lived in communities of Christian hermits.

But Saint Paul's teachings regarding community as the body of Christ reflect the widely held belief that the *parousa* ("presence" or "arrival")—the hope that the risen Christ's return to the world to judge the living and the dead—was imminent. But after Saint Paul's execution in Rome, Christians began to think of the *parousa* as delayed, as a reality that would only come to pass in a future of God's choosing. At this point, as Christian communities began to populate all areas of the Roman Empire, the inclusive character of the earliest Christian communities became more exclusive as doctrinal commitments and moral standards became the standard for membership. In other words, doctrinal commitments and moral lifestyles were gradually established as criteria for inclusion as the various Christian communities became increasingly institutionalized.

Thus the struggle to figure out how to create a *koinonia* that is "the body of Christ" within ever-changing, rough-and-tumble cultural and historical contexts has been ongoing for two thousand years. Since the fourth century, theological exclusivism has more often than not defined membership in Christian communities. Inclusion in the Christian community meant acceptance of "orthodox" doctrinal formulations and practices, which is in contradiction to the historical Jesus's instructions to his disciples recorded in Mark as well as Saint Paul's inclusive view of incorporation within the "body of Christ." Throughout Christian history, great theological debates

3. See Col 1:24.
4. See 1 Cor 12:12–27.

about "community" and who is and who is not a member were often a source of violent injustice against those deemed "heretical," particularly Jews. Non-Christians were, and often are today, treated as objects of conversion to the "only way of salvation." At times, the Ways of non-Christians were treated with fear, as in he case of Islam, or were simply viewed as ways of superstition. No religious Way has spilled more non-Christian blood than this sort of Christian imperialism.

Of course today as more and more Christians are taking religious pluralism seriously and are anxious for dialogical engagement with religious Ways other than their own, the willingness *not* to employ rigid doctrinal and moral norms and liturgical practices as conditions for inclusion in "the body of Christ" is evolving. A list of such groups would include Quakers and Mennonites. Subcommunities within larger Protestant denominations and Roman Catholicism are also more inclusive communities than the wider denominations of which they are a part. Examples abound: Lutherans, Methodists, Presbyterians, American Baptists, and Catholics mostly associated with monastic communities in dialogue with Protestant and non-Christian Ways. "Progressive" Christian communities populate many mainline Christian denominations as subcommunities. These communities are "open and reconciling" communities where "all are welcome" regardless of past denominational affiliation, gender identity, ethnicity, or sexual orientation. "All are welcome" is the slogan of such Christian communities and reflects a desire to embody in the present the meaning or *koinonia* in the historical Jesus's instructions recorded in Mark 9:38.

THE ISLAMIC WAY

The monotheistic religious Ways that originated in the Middle East were focused on establishing compassionate, just communal structures of existence. Community established on compassion and justice was the central focus of Islam from the very beginning of its history. If my reading of the Qur'an is accurate, a communal way of life evolves from surrendering one's will to God's will. In other words, surrendering to God demands struggling to create just, compassionate community that is inclusive of all human beings as well as of the natural processes nurturing human existence.

Accordingly, the Qur'an and the Sunna specifically prohibit race, nationality, occupation, kinship relationships, or special interests from defining principles of community. In Islamic understanding, community is not named after a founder, leader, or a particular historical event. Community transcends national and political boundaries. Community is "the House of Islam," or *dar al-ʿīslām*, composed of all who surrender to the will of God. The act of surrendering to God's will that human beings live justly and compassionately in community is what makes one a *Muslim*, meaning "he or she who surrenders to God's will." It is the only requirement for inclusion in the House of Islam. As the Qur'an states, "Let there be community (*ummah*) among you, advocating what is good, demanding what is right, and eradicating what is wrong. These are successful" (3:104).

Of course Muslims confront the same struggle that participants in all religious Ways face. It is one thing to surrender to God's will and live in community with other Muslims, but the question is always how to do so in the ever-changing conditions of actual historical existence. How should Muslims interact with non-Muslims? How should Jewish and Christians interact with non-Jews and non-Christians? Most often, communal relations have been exclusivist; this in turn has spawned horrible violence, as evinced by Pope Urban II's call to retake Jerusalem from Muslims in 1095. This invasion, known as the First Crusade, initiated a number of crusades that finally ended in 1291. Further, Muslims did not relate compassionately or justly to non-Muslim communities in India and other regions of South Asia. My point is, in similarity with followers of all religious Ways, Muslims still struggle to live up to the Qur'an's communal standards.

THE HINDU WAY

Measured by the call for just, compassionate community in the Israelite and Judahite prophetic tradition, the Hindu Way seems at first glance to have little interest in community. Early in the religious history of India, community was structured around a system of four castes. Society during the Vedic Period (ca. 1750–500 BCE) was patriarchal and patrilineal. Marriage and childbearing were especially important to maintain male lineage. The institution of marriage was important, and different types of marriages—monogamy, polygyny, and polyandry—are mentioned in the *Rigveda*. All priests, warriors, and tribal chiefs were men, and descent was always through the male line. In other parts of society, women had no public authority but were able to influence affairs within their own home. Women were to remain subject to the guidance of the males in their lives—first father, then husband, then son. These distinct gender roles may have contributed to the social stratification of the caste system.

The classical Hindu caste system most probably originated in the well-defined social orders of the Indo-Aryans of the Vedic Period. The Rigveda, one of the sacred canonical texts of the Vedas, points to various deities as the origin of the caste system. The castes were a form of social stratification characterized by the hereditary transmission of lifestyle, occupation, ritual status, and social status. These social distinctions may have been more fluid in ancient Aryan civilizations than in modern India, where sociologists are currently seeing that intercaste marriage and interactions are becoming more fluid and less rigid.

The castes or *varnas* (meaning "color") enforced social divisions that are still a living reality of traditional Hindu communal life in India, but not so much outside India. By around 1000 BCE, the Indo-Aryans developed four main caste distinctions: (1) Brahmins, or priests; (2) Kshatriyas, or kings, rulers, or warriors; (3) Vaishyas, or agriculturalists, artisans, and merchants; and (4) Sudras, or service providers. Each caste was divided into *jatis* ("subcastes"), which identified the individual's occupation and imposed marriage restrictions. Marriage was only permitted between members of the same *jati* (or two that were very close). Both *varnas* and *jatis* determined a person's purity level. Members of higher *varnas* or *jatis* had higher purity status and if contaminated (even by touch) by members of lower social groups, they would have to undergo purification rites.

As the Aryans expanded their influence, the newly conquered groups were assimilated into society as new groups formed below the Sudras, outside of the caste system. These outcasts were called the "untouchables" because they performed the least desirable activities, such as dealing with dead bodies, cleaning toilets and washrooms, and tanning leather.

The caste system has survived for over two millennia, becoming one of the basic features defining how conservative Hindus think of "community" even as protest movements against caste have always existed throughout India's history, most notably from Buddhists and Jains, but also from orthodox Hindus as well. Although India's constitution formally abolished the caste system in 1950 because of criticism from such eminent Hindus as Mahatma Gandhi and Vinoba Bhave, and because of the influence of Islam, and the Western democratic influences on the Indian Independence Movement, some Hindus continue to participate in the caste system.

THE BUDDHIST WAY

The Buddhist Way began as a monastic community supported by lay followers. In fact, the Buddhist Way was the first religious tradition to emphasize cloistered life apart from worldly or secular affairs. Community (*saṃgha*) is one of the "three jewels" of Buddhism, the others being the historical Buddha and the Dharma, meaning something like the "teachings of the Buddha." "Taking refuge in the Three Jewels" initiates persons into membership in the Samgha. But originally, *saṃgha* referred only to the community of monks and still does in South Asian Theravada (School of the Elders) tradition. But the Saṃgha evolved to include lay participants in the schools of East Asian Buddhist tradition collectively referred to as Mahayana or "Great Vehicle." Most Buddhists and non-Buddhists today think of the Saṃgha as inclusive of all practicing Buddhists.

As the Buddhist Way gradually shifted from groups of wandering monks, and later monks living in monasteries, and in Mahayana tradition nuns living in their own convents, monastic structures were integrated into Buddhist lay communities. Indeed, monasteries, convents, and temples were dependent upon lay Buddhist support, apart from which they could not have survived.

Among the earliest known Buddhist activities that included lay participants were those that occurred at sites thought to contain relics of the historical Buddha's body, called *stupas*. These areas became sites for pilgrimage and joyous festivals involving locals and visitors in multifaceted events that at once entertained, informed, and generated merit. Gradually, more and more monasteries were built, not only near these stupa sites but also in many other locations in South Asia. As Buddhism was established throughout Asia, the monastery became its fundamental institutional structure.

Although Buddhism was initially disruptive of conventional social structures by drawing men and women away from their social obligations, it soon became a productive contributor to local communities. A Buddhist monastery provided the lay community with religious rituals, spiritual guidance, places of worship, and a means of earning merit. Buddhist laypersons sometimes formed social clubs that arranged pilgrimages, met to chant or copy sutras, or carried out social projects for the benefit of the community. Thereby, the Buddhist Way became known as a source of practical benefits for its followers. This practice grew from the texts and stories about the bodhisattvas, who were objects of devotion that might help laypersons attain Awakening and were also thought to protect people from dangers such as fires or floods. One might visit a monastery to ask for healing for a loved one, for help finding a marriage partner or employment, to ensure the birth of a healthy child or grandchild, or to ask for success in examinations or a myriad other things that would benefit one's life.

Monasteries also served social needs by determining auspicious dates for weddings, telling fortunes, and, in some cases, providing political clout and mediation between local communities and powerful, sometimes oppressive politicians. Monasteries served as inns, offering beds and meals for travelers. Some provided social services such as schools, orphanages, health-care facilities, food and shelter for the homeless, facilities for the elderly, and even animal rescue. Occasionally they carried out public works projects such as building roads and bridges, digging wells and planting trees along travel routes, deepening river channels, or creating reservoirs to provide lay communities with fresh water. This tradition is carried out in new ways to confront the political and social issues of contemporary life in a movement called Socially Engaged Buddhism.

Socially Engaged Buddhism, also known as Engaged Buddhism, is not a sectarian movement but rather a specific movement inclusive of a number of Buddhist communities. Engaged Buddhism is a cross-denominational movement that involves both the lay community and monks and nuns and also includes Western converts as well as Eastern Buddhists. While maintaining the Buddhist emphasis on inward spiritual growth through meditation, Engaged Buddhists also seek to confront suffering and oppression through political and social reform. Engaged Buddhists look to the compassionate bodhisattvas of Mahayana Buddhism, who postponed their own enlightenment to assist others, as the ideal. As I previously noted, the term "Engaged Buddhism" was coined by Thích Nhất Hạnh in 1963, at a time when his country was ravaged by the Vietnam War. Thích Nhất Hạnh now lives in exile in a monastery in France. He remains the most revered leader of the movement and has established the "Order of Interbeing" to promote social causes.

In his book *Interbeing*, Nhất Hạnh lays out the following fourteen Precepts of Engaged Buddhism, which I have paraphrased below.[5] These principles emphasize social change as beginning within oneself.

---

5. Nhất Hạnh, *Interbeing*, 17–49.

1. Do not be idolatrous about or bound to any doctrine, theory, or ideology, even Buddhist ones.

2. Do not think the knowledge you presently possess is changeless, absolute truth. Avoid being narrow-minded and bound to present views. Learn and practice nonattachment from views in order to be open to receive others' viewpoints.

3. Do not force others, including children, by any means whatsoever, to adopt your views, whether by authority, threat, money, propaganda, or even education. However, through compassionate dialogue, help others renounce fanaticism and narrow-mindedness.

4. Do not avoid suffering or close your eyes before suffering. Do not lose awareness of the existence of suffering in the life of the world. Find ways to be with those who are suffering, including personal contact, visits, images and sounds. By such means, awaken yourself and others to the reality of suffering in the world.

5. Do not accumulate wealth while millions are hungry. Do not take as the aim of your life fame, profit, wealth, or sensual pleasure. Live simply and share time, energy, and material resources with those who are in need.

6. Do not maintain anger or hatred. Learn to penetrate and transform them when they are still seeds in your consciousness. As soon as they arise, turn your attention to your breath in order to see and understand the nature of your hatred.

7. Do not lose yourself in dispersion and in your surroundings. Practice mindful breathing to come back to what is happening in the present moment. Be in touch with what is wondrous, refreshing, and healing both inside and around you.

8. Do not utter words that can create discord and cause the community to break. Make every effort to reconcile and resolve all conflicts, however small.

9. Do not say untruthful things for the sake of personal interest or to impress people. Do not utter words that cause division and hatred. Do not spread news that you do not know to be certain. Do not criticize or condemn things of which you are not sure. Always speak truthfully and constructively. Have the courage to speak out about situations of injustice, even when doing so may threaten your own safety.

10. Do not use the Buddhist community for personal gain or profit, or transform your community into a political party. A religious community, however, should take a clear stand against oppression and injustice and should strive to change the situation without engaging in partisan conflicts.

11. Do not live with a vocation that is harmful to humans and nature. Do not invest in companies that deprive others of their chance to live. Select a vocation that helps realize your ideal of compassion.

12. Do not kill. Do not let others kill. Find whatever means possible to protect life and prevent war.

13. Possess nothing that should belong to others. Respect the property of others, but prevent others from profiting from human suffering or the suffering of other species on Earth.

14. Do not mistreat your body. Learn to handle it with respect. Do not look on your body as only an instrument. Preserve vital energies (sexual, breath, spirit) for the realization of the Way. For brothers and sisters who are not monks and nuns, sexual expression should not take place without love and commitment. In sexual relations, be aware of future suffering that may be caused. To preserve the happiness of others, respect the rights and commitments of others. Be fully aware of the responsibility of bringing new lives into the world.

THE CHINESE WAY

A common view of the Daoist Way is that it encourages people to live with detachment and calm. The guiding assumption at work in this interpretation is that Daoists live in monastic-like communities in separation from the majority of human beings, not in the sense of being antisocial or asocial but rather suprasocial and often simply different. Daoists, according to this view, neither criticize society nor support it by working for social change but go along with the rhythms of natural forces as the Dao moves through them, without much concern for rules and proprieties of conduct, which they leave to the followers of the Confucian Way.

But contrary to this rather stereotypical understanding, Daoists through the ages have developed various forms of community and proposed numerous sets of behavioral guidelines and texts on ethical considerations.

Like the ancient pre-Daoist and pre-Confucian philosophers, who are well-known for the moral dimension of their teachings, the Daoist Way also focused on ethics: on the personal values of the individual and on the communal norms and social values for organizing communal life. These ranged from basic moral rules against killing, stealing, lying, and sexual misconduct to suggestions for altruistic thinking and models of social interaction to behavioral details about how to bow, eat, and wash. Finally the emphasis on ethics also involved teaching people to live in imitation of the Dao itself. About eighty texts in the Daoist canon and their supplements describe such guidelines and present the ethical and communal principles of the Daoist Way. These texts document just to what degree Daoist realization is based on how one lives one's life in interaction with the community—family, religious group, monastery, state, and cosmos. Ethics and morality, as well as the creation of community, emerge as central in the Daoist Way.

The Daoist Way and the Confucian Way share a common worldview centered on the Dao, but interpreted differently how human beings ought to live in accordance with the Dao's ever-shifting rhythms of balance and harmony.[6] But the view that Daoists formed their own communities isolated from struggle to build communal political structures, while Confucians were only engaged with the politics of statecraft unconcerned with nature, is a bit of a caricature. The Chinese people, taking to heart the necessity of balancing *yin* and *yang*, sought to structure communal life by seeking balance between the *yin* tradition of the Daoist Way and the *yang* way of life specified by Confucius and later philosophers.

The primary goal of Confucian practice is raising awareness of *ren* or "humanity," the theme around which all Confucian teaching revolves. A Superior Human Being or "Sage" is one who has trained himself or herself to apprehend the common humanity incarnated in all human beings, which defines human beings as "human" and distinguishes human beings from other sentient beings. *Ren* is the expression of the *dao* working in human life. That is, like all things and events in nature, human beings are a specific reflection of the *dao*. So "humanity" is the realization that all human beings are particular, interconnected, and interdependent expressions of the *dao* and therefore should be interacted with accordingly. Realization of "humanity" is want transforms an ordinary person into a "superior human being" or "sage" (*chün tzu*) who has learned to live within the ever-shifting

---

6. Ingram, *Passing Over and Returning*, ch. 6.

points of balance between extremes, like a good host who is able to put others at their ease by knowing what to do in any political or social situation—and doing it. But how does one become a superior human being? The Chinese Way's answer originates with Confucius. One may gradually become a sage by means of four interrelated disciplines: (1) *ri* or "filial piety"; (2) *meng*, literally "names" but better understood as "the rectification of names"; (3) *zhongwen* or Doctrine of the Mean; (4) *wen* or "culture."

"Filial piety" refers to the concrete ethical expression of "humanity" in communal interaction through the "five filial relationships": (1) father and son; (2) husband and wife; (3) elder brother and junior brother; (4) elder friend and junior friend; (5) ruler and subject. Note that what defines the superior/inferior relationship is age and gender. Generally, Confucius and subsequent Chinese tradition taught that males are socially superior to females (which means that Confucian communities are patriarchal social structures), and older people are socially superior to younger people. Also the superior side of each relationship is identified as *yang*, and the inferior side as *yin*, a notion *not* found in the Daoist Way.

So fathers (or mothers or parents) have obligations to their children that must be fulfilled, which are created by the very fact of having children: basically to see to the protection, education, and welfare of their children. Given what sons particularly, but also daughters, receive from fathers or parents, children have the obligation to be obedient to parental authority, even after they have become married adults raising their own families. Likewise elder brothers have obligations to look after and be positive examples for a younger brother's conduct, while younger brothers have the duty to be obedient and respectful to elder brothers. The same is true for the elder sister / younger sister relationship. Similar superior/inferior responsibilities occur in the relationship between elder friends and junior friends. Finally, for what rulers do to bring order and stability into a community by keeping at bay forces that cause social chaos and disharmony, subjects owe rulers submission and obedience. Since everyone is in either a superior or an inferior social relationship to someone in every human community—because one is always older or younger or male or female—knowing *what* to do and *publicly* doing it through a display of socially constructed rituals creates harmony that expresses the balance of *yin* and *yang* communal forces. Thus superior human beings teach "humanity" to others by their public display of filial piety.

"The Rectification of names" refers to the disciplined use of language. Our sense of reality, the way things really are, is largely a linguistic construction. For example, the meanings of realities such as "death," "life," "old age," or "Dao" are what human beings construct through linguistic reflection on experience. Language is the primary means by which we understand anything, but at the same time language is the primary means by which we falsify reality—for ourselves and for others. Consequently, in similarity with all religious Ways, the disciplined use of language as a means of apprehending truth and communicating it is of utmost importance both for individuals and for the welfare of community. Communal relations between individuals absolutely require such discipline. For before a father can fulfill his filial responsibilities to his children, he must know the meaning of the word "father." Before a son or a daughter can fulfill their filial responsibilities to their parents, they must know the meaning of "son" or "daughter." Knowing the filial obligations owed to a ruler and knowing the obligations a ruler owes to his or her subjects are absolutely required to maintain the social and political order in the state.

Accordingly, Superior Human Beings use words carefully and clearly, with as much univocal meaning as possible, without equivocation, so that when they speak there is no ambiguity or misunderstanding about what is meant. Communal relationships are all too often broken because one party misunderstands what the other party means. Furthermore, relationships between communities and states depend on linguistic clarity. The chaos of war has often originated from such misunderstandings.

"The doctrine of the mean" points to how human beings should live in accordance with the *dao*. Superior Human Beings do nothing in excess by ceaselessly seeking the proper balance between extremes in every communal situation while taking action accordingly. That is, the communal vision of the Chinese Way—which seeks to harmonize the Confucian Way and the Daoist Way—counsels living in balance between *yin* and *yang* through communal engagement with the hard realities established by social and political obligations for the creation of compassionate and just communal structures that help human beings live together harmoniously for the common good.

Finally, Superior Human Beings practice "culture," sometimes called "the arts of peace." To practice "culture" may sound a bit odd to Western ears. But it helps to keep in mind what the word *wen* or "culture" means. There are specific kinds of human creativity found nowhere else in nature that point to what all human beings share in common as individual reflections of the *dao*. This common reality shared by all human beings is called *ren*, "humanity." *Wen* or "culture" is the objective expression of that which publicly reveals *ren*, which constitutes us as "human": art, literature, philosophical reflection, calligraphy, technology, "science." Study of culture and the practice of culture reveal what is unique about human nature and the construction of human community. Thus Superior Human Beings, through the study and practice of culture, become, in the words of Mengzi in the *Book of Mengzi*, "superior within and kingly without."[7]

But the Chinese Way is quite realistic. Becoming a Superior Human Being is a lifelong process, which means building compassionate and just human community is a lifelong process as well. Actually creating just, compassionate, communal structures within the processes of historical existence is beyond final attainment. But nevertheless, as the Chinese Way maintains, the *struggle* to attain it must continue.

CONCLUDING OBSERVATIONS

Apart from concrete embodiment, compassion, justice, and community are abstract universals that Alfred North Whitehead referred to as "eternal objects."[8] As ideals, eternal objects having no existence apart from the prehension of actual human beings—that is, until incorporated into the actual experience of human beings, who in turn are undergoing ceaseless processes of interdependent becoming. But eternal objects in and of themselves are empty other than as objects to be prehended—that is, until concretized (but never fully) within the collective experiences of actual human beings. Thus when the prophet Micah proclaimed, "You have been told what is good," he was pointing to universal ideals—eternal objects—that in his time and place he negatively prehended as nonexistent in any collection of human beings, and especially absent in eighth-century-BCE Israelite and

---

7. Chan, *Sourcebook in Chinese Philosophy*, 67. Mengzi lived in the fourth century BCE. His interpretation of the sayings of Confucius collected in the *Analects* became the foundation of the "orthodox" tradition of the Confucian Way from the Han Dynasty in the fourth century until the establishment of Mao Tse Tung's Marxist-Communist regime in 1948.

8. See Whitehead, *Process and Reality*, 163–66.

Judean society—a society that should have known better given its claim of covenant relationship with Yahweh.

So in no place or time inhibited by human beings has just, compassionate community been fully attained. At no place and time have human beings fully lived compassionately and justly in community, although certainly there have been times in history when human beings have, if only partially, glimpsed these realities more fully than their contemporaries. One thinks of the early Jesus movement and Jewish communities within the Roman Empire; or of the small group of people who followed Saint Francis's itinerant wanderings in thirteenth-century Spain, and of the later spiritual Franciscans; or of the Quaker movement initiated by George Fox; and in the twenty-first century one thinks of numerous Roman Catholic and Greek Orthodox monastic communities or of Buddhist Social Engagement or of Mahatma Gandhi's independence movement or of socially engaged Buddhists such as Thích Nhất Hạnh.

The struggle continues to this day, at least for some societies of human beings attempting to build communities sensitive to the common good of their members. But unjust forces actively resist and counter the creation of compassionate, just communal structures everywhere on planet Earth. So while most human beings to varying degrees continue to "know what is good," the creation of compassionate, just community is as elusive as it was in the past—perhaps more so. A central reason this is so in our time has to do with our dominant economic system known as free-market capitalism and its consequences, which include environmental damage caused by global warming, rising poverty rates, and increasing numbers of homeless and hungry human beings throughout the world. And most certainly one of the forces that resists community, the state of continual warfare caused by America in places like Vietnam from 1965 until 1975 and currently in the Middle East, results from free-market economic policies, most particularly the desire for profits by American and British oil corporations and arms dealers—who seem to be writing the laws relative to their interests.

So the remaining chapters of this book will focus on ways that the current economic system might be reformed in light of climate change. But there exists no optimism in my conclusions, for even if the revisions I suggest to reform the current economic system governing all human beings everywhere at all times on this planet were to actually take place, it will probably be too late to avoid the global catastrophes I think are afoot in the near future. I am certainly not a prophet, but I think there are economic signs of the times to which human beings had better pay attention before it's too late.

4

"The Invisible Hand"

I must to be honest. I am *not* an economist. Nor have I ever wanted to be an economist. The source of my disinterest began during my first and only undergraduate "Introduction to Economics" course at Chapman University. It wasn't that I found the history of economics beginning with Adam Smith or economic theories uninteresting. The source of my disinterest was the course's instructor, who claimed that the differences between the plurality of human cultures, ethnicities, religious identities, and gender could be "smoothed out" by capitalism, which would eventually meet the needs of all persons who were "willing to work hard." "There is no other economic system like it anywhere" was the mantra he chanted at the beginning of each class session. He demonstrated his faith by reducing all things economic to equations that rivaled the complexity of the equations I had studied in courses in physics and mathematics. Indeed, economics seemed to me a "dreary science."

I was a working-class kid from Santa Monica, California, and this seemed a pipe dream. My father and mother, who grew up during the Great Depression, knew by experience and hard work that they were locked into an economic system that pressed heavy as my father worked sixteen-hour days painting cars in numerous body shops. He loved reading in his spare time, and he was one of the most intelligent people I have known. His work ethic taught me much about economic theory that my professor left out of his introductory course, particularly the absurdity of reducing human beings to abstract mathematical formulations. So at the time, economic reductionism seemed stupid to me. What I took away from this course was a deep suspicion of what I later learned Alfred North Whitehead called "the fallacy of misplaced concreteness."

Even so, I now understand that we ignore the economic principles underlying capitalism at our collective peril, particularly free-market capitalism. I came to this realization after reading John B. Cobb's, book *For the Common Good*. Along with his coauthor, economist Herman E. Daly, Cobb engaged in a detailed theological critique of current economic theory, particularly postmodern free-market capitalism. Daly and Cobb pointed me towards other economists such as Gordon Douglas and Lee MacDonald as well as progressive Christian theologians critical of economic theory, particularly Ward McAfee.[1] I have come to agree with their conclusions that the current economic domination system to which most human beings have no choice but to surrender must be drastically challenged and changed "for the common good."[2] Not doing so will result in unimaginable worldwide poverty, social chaos, and ecological destruction that even now threatens numerous species along with our own. We must challenge the dominance of free-market capitalism "before it's too late." The only question is, how?

Since the issues are global, the religious Ways of the world are a collective resource for socially engaging and perhaps transforming capitalism into a less oppressive and more just economic system trough the inclusion of socialist economic principles.[3] In what follows, I shall draw upon the vision of Whiteheadian process theology as I show how the religious Ways I have surveyed in this book have critiqued, and in some cases have tried to resist, the disruptive economic forces now threatening human community and the environment. But first, the economic principles and practices of free-market capitalism must be clearly described and analyzed. Since I am not a practicing economist I am relying on the work of the economists mentioned above, plus others whom I will identify as the my discussion unfolds.

1. See Daly and Cobb, *For the Common Good*; Douglas and McAffe, "Consumerism"; Douglas, "Poisonous Inequality."

2. See Cobb, *Sustaining the Common Good*; and Cobb, *Sustainability*.

3. See Clayton and Heinzekehr, *Organic Marxism*, chs. 12–14.

If there is a single text that serves as "scripture" for capitalist economic theory it is Adam Smith's *An Inquiry Into the Nature and Causes of the Wealth of Nations*, also shortened to *The Wealth of Nations*.[4] First published in 1776, Smith argues that wealth grows more rapidly if governmental policies do not favor competing economic interests, as they in fact did when Smith was writing his book. Smith observed that early capitalism flourished in this environment because it brought new wealth to a new class of entrepreneurs as well as greater wealth to the English nobility. But he predicted that wealth would increase more rapidly for more classes of people if capitalism became free from restrictions, particularly those of the Navigation Acts of his day, which favored one country's products over those of another. In other words, free trade would engender increased production of goods and more wealth to purchase goods.

Smith built upon economic ideas that existed before him, being at times more like a realizer than an innovator. In particular, his theory built on the economic implications of the philosophy of René Descartes and the science of Isaac Newton. So while the specific ideas of Smith's economic philosophy were not original in his time, he was the first to compile and publish capitalist economic ideas in a systematic way. For this reason, among others, he was responsible for popularizing many of the ideas and assumptions that underpinned his economic philosophy now known as "classical economics." Later economists built on the *Wealth of Nations* as classical economics became the dominant tradition of economic practice during the Great Depression in western Europe, Canada, and the United States. The fundamental ideas of Smith's economic philosophy is classically summed up a French slogan, *laissez-faire*, "let them do" or "let them alone."

Smith's laissez-faire version of capitalism called for minimizing the role of governmental intervention and taxation in "free markets" because an "invisible hand" guides the demand, supply, and flow of goods and services. Smith assumed that human nature is greedy and competitive so that individuals look out only for themselves and those closest to them; the doctrine of original sin in traditional Christian theology articulates a similar assumption about human nature. But—and this is Smith's point—individual self-interest creates the best outcome for everyone. Each individual by looking out for himself or herself inadvertently creates the best economic decisions for all persons. So it is not the benevolence of a butcher or brewer or baker upon which we can depend for food, but their self-interests for

4. Smith, *Wealth of Nations*.

making an economic profit. Furthermore, Smith noted that people would invest their accumulating wealth in enterprises most likely to earn the highest monetary returns.

From Smith's economic theory various theories of capitalism have evolved; his theory finds contemporary expression in free-market capitalism. As Daly and Cobb point out, the paradigm of free-market capitalism focuses on *homo economicus*, which points to this theory's particular assumptions about human nature.[5] First, free-market capitalism pushes Adam Smith's theory to its furthest conclusion in the way it builds on the universal propensity to optimize individual self-interest in competition with other individuals attempting to do the same thing. This is why in the last half of the nineteenth century Charles Darwin's theory of the "survival of the fittest" as the mechanism driving the biological evolution of species was grafted into capitalist economic theory.

Current economic theory typically identifies the intelligent pursuit of individual benefit with "rationality" while concluding that noneconomic modes of behavior are "irrational." These modes of behavior include "other-regarding" behavior and actions directed toward the pubic good. In other words, *homo economicus* is a "pure individual" separated from other "pure individuals." Accordingly, "capitalism" consists of private ownership of the means of production along with the allocation and distribution of the market. Individual maximation of profits by firms and maximation of satisfaction (utility) by consumers provide the motive force, while competition, the existence of many buyers and sellers in the market, provides "the invisible hand" that leads private interest to serve pubic welfare. Contemporary free-market capitalism is Smith's theory of capitalism on steroids.

For Smith, above all the collective role of government must be limited to (1) protecting individual property rights enforced by law, (2) providing "public goods and services" (public monopolies) while prohibiting the formation of private monopolies, (3) maintaining "aggregate demand" at a level that maintains an "acceptable" combination of inflation and unemployment, (4) providing a minimum social-welfare safety net to keep people from destitution, and (5) intervening to correct "externalities," situations in which voluntary exchange between two individuals, although mutually beneficial to them, has important negative effects on third parties.[6]

5. Daly and Cobb, *For the Common Good*, 7.
6. Ibid., 13.

The individualism inherent in free-market capitalism is expressed in the self-interested behavior of individual persons. This system of economics leaves no place for fairness, no place for either malevolence or benevolence, no place for preserving human life, or any other moral concerns. The world that this economic theory pictures is one in which all individuals seek their own good in indifference to the successes or failures of other individuals engaged in the same activity, and, as previously noted, is a prime example of Whitehead's "fallacy of misplaced concreteness" in the way it abstracts from the actual communal character of human existence. Yet in the real world, the biological evidence is in: *self-contained individuals do not exist.* Thus the individualism of current economic theory is profoundly erroneous because human beings are social, that is, are constituted by their relationships to other human beings *and* to the life forms with which we share planet Earth. That human beings are constituted by interdependence with, not separation from, all things and events is clearly a central teaching of the Buddhist Way, but it is also underlies the Israelite and Judahite prophetic tradition, the life and teachings of the historical Jesus, Islam's call for compassionate justice, the Hindu Way, the Chinese Way, and the Jewish and Christian Ways.

In its most basic meaning community assumes that human beings are internally related to one another. That is, their relationships define their identities as persons, which means that any viewpoint that treats persons as self-contained individuals falsifies the real situations in which persons must live. Because interdependence and the virtues of compassionate justice are simply absent from capitalist economic theory and practices, the search for community is simply abandoned as illusionary. What is left is an abstraction called the "invisible hand" creating groups of individuals whose sole purpose is the consumption of the goods and services of the marketplace in isolation from other consuming individuals.

Since consumerism is what energizes free-market capitalism, it is necessary to describe in some detail consumerism's structure of existence. Of all the research I have undertaken on consumerism, the clearest thought I have encountered is Gordon Douglas and Ward McAfee's essay, "Consumerism."[7] Since Douglas, who is a distinguished economist, wrote the section on consumerism in this essay, I shall follow his lead.

Since Adam Smith published *The Wealth of Nations* in 1776, capitalist economic theories are the collective source of the rise of consumerism now

7. Douglas and McAfee, "Consumerism," 55–74.

on a global scale. The logic of economic theory originates in the ideal of permanent scarcity of the means of production—that is, scarcity encountering limitless human needs and wants. Consequently, capitalist theory seeks ways of producing more output from less input, which means that what is demanded is the most efficient allocation of resources and capital. So the quest for efficiency trumps other social goals that are not marketable commodities and have no regularly assigned prices. Thus, the usual question is, how much might a given social improvement damage the production of more goods desired by consumers? Liberals and conservatives alike accept this "efficiency premise" so that economists mostly argue over the best way to achieve more, not whether "more" truly contributes to human well-being.

Douglas lists five factors that drive economic efficiency that have profound negative effects on the global environment. The first is advertising. The manipulation of the desire to own possessions, in other words "consumerism," is the engine driving free-market capitalism. Essentially, advertising is "information" about available goods and services. Since the power of advertising is propaganda (the manipulation of human beliefs and desires), "information" must be taken with a grain of salt. Advertising is a central feature of consumerism. Although advertising of some sort has been part of all societies, it has evolved far more fully in western Europe and North America beginning during World War I. Since that time a profound cultural shift has occurred as commercial appeals now hammer human beings to possess all the goods and services they can because, as the character Gordon Gekko declares in the 1987 film *Wall Street,* "greed is good." In the process rational consumer decisions are weakened.

The banalities of contemporary advertising campaigns are now as commonplace as the material delights of goods and services that are displayed in ways to make them seem irresistible. But of course not all can afford to participate in a continuous spending spree to own and possess things, and the negative effect on both the rich and the poor is profound, as people gradually come to the painful conclusion that no one can really own anything. The truth is, the more we think we own anything, that thing winds up owing us—and the buying spree continues. Consumerism is thus the "religion" practiced by most persons in western Europe and North America.

The second characteristic of consumerism is planned obsolescence. Advertising makes it possible to sell new goods to replace old goods that still function well. Models of products are constantly changing, requiring ever more purchasing. Planned obsolescence is incorporated into specific products. In other words, planned obsolescence is a built-in feature of free-market capitalism; without it, the current economic system would simply break down.

Third, an underlying anxiety supports current economic practice. With the emergence of the nuclear age in 1945 after the United States bombed Hiroshima and Nagasaki, the terror of "mutually assured destruction" of the Cold War between the United States and the Soviet Union became a hard reality for the next thirty years. The global anxiety of this period led to a buying spree as if there were no tomorrow—because of the possibility there might actually be no tomorrow. Consumerism evolved into a diversion from the fear of nuclear war. Patterns of consumption and pleasure seeking became part of European and North American culture, with each new generation of human beings trying to outdo its predecessors.

The fourth characteristic of current economic practice is easy credit. The evolving U.S. credit system that emerged after the end of World War II allowed Americans to borrow in order to satisfy their collective desires for more and more goods and services. A large percentage of Americans borrow against their savings, using credit cards and lines of bank credit that steadily diminish the accumulated equity they own in their homes. Credit cards offer everything from frequent-flyer miles to low introductory interest rates to low minimum payments. It wasn't long before credit cards from every large bank in America were pushing "easy payment" cards on adolescents in order to draw them into a culture of consumerist debt. Today, less than one-third of Americans avoid interest payments by paying off their monthly credit card balances.[8]

Finally, the consumerist engine that drives free-market capitalism is its widespread acceptance. It is the "religion" practiced by most contemporary North Americans and western Europeans, and it is rapidly spreading throughout the rest of the world. This has been so since the deep economic misery most persons experienced with the Great Depression. The prosperity initiated by the New Deal allowed the creation of a welfare state of sorts, although current welfare programs for the poor and homeless are now under serious attack in the United States by the Republican Party. Since

---

8. Ibid., 60.

no one wishes to return to another Great Depression, consumers buy and sell anything and everything in the hope of securing permanent economic well-being and security for themselves and their families. It is, appropriating a phrase from theologian Paul Tillich, the "ultimate concern" of most human beings on the planet.

But the most fundamental and globally dangerous consequence of consumerism is its effects on communal and environmental interrelationships. Global capitalism is constructed on a set of circular expectations economists call the "industrial package." The "industrial package" includes several components: "(1) output per worker (worker productivity) will advance through (2) technological and institutional innovations based on (3) increased energy use and material inputs, accompanied by (4) higher purchasing power supported by (5) a consumerist mentality that assures that the things produced will be purchased." [9]

Here lies the source of consumerism's damage to the environment. The high consumption rates demanded by current economic models are not environmentally sustainable. Consumption is rising much faster than improvements in material usage or in the energy use required for material output (point 3), which engenders dangerous implications for the natural world. Most economists argue that technological innovations and improvements in management will increase the productivity of energy per unit of output over time (point 2). But environmentalists doubt this is the case because it does not seem possible based on any known technology for the people of China or India or other countries with overpopulation issues.[10] In other words, as consumerism generates poverty worldwide, it is also one of the sources behind global warming that is now destroying the ecosystems upon which all life depends for survival.

Making a similar point, John Cobb writes that from a Christian viewpoint the effects of the free-market capitalist model are severely damaging: "if the policies of this model are continued much longer they will be disastrous."[11] This is so because (1) current economic models describe how production increases with the increase of growth and specialization without taking into account the input of raw materials in the production process or the emissions of waste into the environment, and (2) the scale of human production is so great that these damaging effects are global and

9. Ibid., 62.

10. McKibben, *Eaarth*, ch. 1.

11. Cobb, *Sustaining the Common Good*, 8.

catastrophic. The economic goal of increasing production even more, the central goal of free-market capitalist models, encourages the further and intensified degradation of the environment.

The problem is that human beings have been transformed into *homo economicus*, to borrow a phrase from Cobb.[12] This transformation began with Adam Smith's publication of *The Wealth of Nations*. While Smith did not intentionally incorporate Calvinist theology into his economic theory, he agreed with Calvin that human beings are only motivated by greed and self-interested individualism when it comes to economic relationships. Consumerism as the engine driving contemporary economic practice is the end result. But each of the religious Ways described in this book more realistically understand human beings as persons-in-community rather than consuming individuals isolated from other consuming individuals in the marketplace. Current economic practices break communities apart in the name of individualized self-interest while dismissing the principle of compassionate justice as either a value of the past or an illusion.

Accordingly, from the perspectives of the Jewish, Christian, Muslim, Hindu, Buddhist, and Chinese Ways—each according to their own distinctive principles—participation in a healthy community is more important to human well-being than the consumption of goods and services beyond what is necessary for biological health. Since the notion of community is completely absent from contemporary economic theory, economic practice has been, and still is, engaged in a global assault on human communities and the glue that holds communities together, compassion and justice. And since we live in an interdependent world in which human beings are part of the natural order, contemporary economic practices are in process of destroying the environment upon which human community depends. That is to say, the environmental community in which human beings and all life on planet Earth are a part is now under profound threat.

The most graphic examples of this assault on community are the so-called Second and Third Worlds. American developmental models foisted on less industrialized nations treated the traditional communities of these nations as major economic obstacles. For example, the traditional ideals of community nurtured values that made workers hesitant to leave their communities even when high wages were available is a major obstacle that corporations transplanted to these nations had to overcome in the name of profits for stockholders who did not live in these nations. Whenever

---

12. Ibid., 8–9.

and wherever an opportunity for increasing profits is found, the current economic systems encourages corporations to move. While the mobility of capital keeps the corporate economy growing, the destructive effects of factory closings on the communities where they had been located have no role in corporate economic calculations because these effects are viewed as a sentimentality that limits the ultimate corporate goal: economic growth.

Corporate destruction of community is also rampant in the United States. Since World War II one focus of free-market capitalist theory has been agriculture, which has resulted in increased agricultural productivity—as measured by total product divided by hours of human labor. This productivity has been achieved by mechanization, increased use of pesticides and herbicides, and a decrease in human labor. As corporate farms increasingly have replaced family farms, thousands of rural communities have been destroyed. This in turn has resulted in mass migrations into inner cities and their suburbs, where the quality of community is inferior to what was destroyed in the countryside. This situation is even worse in Second and Third World nations.

Cobb rightly concludes that this shift to corporate agriculture demonstrates the intimate connections between human activity and changes in the biosphere upon which all living things depend. It is certainly true that small family farms are capable of degrading the earth. Still, in many parts of the world small family farms have operated for millennia. So it is possible to reform agriculture, particularly in North America and Europe, along sustainable lines. Furthermore, the sources of pollution in the practices of family farming are small in comparison to the factory farming rapidly replacing them.

Since the engine driving free-market capitalism is consumerism—which is now devouring the resources of planet Earth necessary to sustain life—it will be useful to summarize the traditional economic views of the religious Ways considered in this book.

THE JEWISH WAY

The Torah not only commands Jews to give to the poor but also to advocate on their behalf, or as Prov 31:9 has it, "speak up, judge righteously, champion the poor and the needy." This is the heart of community: helping fellow human beings in need is not merely a matter of charity but of compassionate justice. So Jewish faith and practice center on the moral obligation to advocate for children, the elderly, the poor, the disenfranchised, the sick, the disabled, and the "stranger among us," not on corporate profits or a life of consumption.

The Tanak also offers a detailed description of one of the world's earliest social welfare systems as a means of creating human community reflective of an agriculturally based society. Two examples instruct landowners, for example, to leave the corners of fields and the gleanings of harvest to the poor (Lev 19:9) and to lend to people whatever they need (Deut 7–11). Jewish history also provides examples for helping the needy. During Talmudic times, much community building was done though tax-financed, community-run programs that provided for the poor, the hungry, the ill, and children—which established the precedents for the types of social engagement programs Reform Judaism preserves today.

The Torah and Jewish tradition explicitly focus on feeding the hungry. The Talmud also explains that each Jewish community must establish a public fund to provide food for the hungry, and the Jewish sages explained that feeding the hungry is one of the most important obligations human beings have: "When you are asked in the world to come, 'What was your work?' and you answer: 'I fed the hungry,' you will be told: 'This is the gate of the Lord, enter into it, you who have fed the hungry'" (Midrash to Ps 118:17).[13] In Isa 58:7, God commands Jews to "share your bread with the hungry and bring the homeless poor into your house." Guided by these values, all contemporary traditions of the Jewish Way—Orthodox, Conservative, Reform—support a variety of antihunger programs in the United States and throughout the world, including emergency assistance programs, food banks, food stamps, and child nutrition programs.

In particular, the Israelite and Judahite prophets exhorted their hearers to follow a tradition of hospitality inclusive of Israelites, Judeans, *and* Gentiles. According to one midrash, Abraham is judged to be greater than Job because while the Job "opened his doors to the road" (Job 31:32), Abraham left his tent to welcome strangers (Gen 18:1–8). More recent Jewish history, with its exiles and expulsions, is a powerful reminder of the obligation to provide for those having no protection. Accordingly, the Jewish Way generally supports public policies that address and answer the grievous need for low-cost housing among people in low-income categories in order to improve the quality and availability of housing for impoverished families.

---

13. https://www.on1foot.org/text/midrash-psalms-11817, Danny Siegel, trans.

The Torah also addresses labor issues by emphasizing that laborers be treated justly; see, for example, Deut 24:14–15: "You shall not withhold the wages of poor and needy laborers, whether other Israelites or aliens who reside in your land and towns. You shall pay their wage daily before sunset, because they are poor and their livelihood depends on them." Since the advent of labor unions, many American workers have had a voice in the terms of their own employment. Unions are models of self-sufficiency, allowing workers to stand up to demand their own rights. Unionization has brought real benefits to hardworking Americans in addition to the dignity that comes with workers negotiating as equals with their employers. Jews have an obligation not only to assist the downtrodden but also to help those in need become self-sufficient—an obligation most Jews pursue by promoting the unionization of workers.

In summary, the Jewish Way is deeply committed to achieving a just society grounded in compassion, in which all people can live in dignity and respect. This commitment in today's industrial-based societies includes support and advocacy for employment programs, family planning, social welfare entitlements for public housing, health and legal services, and income-maintenance assistance programs. In the same vein, Jews generally oppose efforts to cut funding for education, job training, food subsidies, and many other social programs that are in danger of losing some, if not all, of their financial resources. Although Jews recognize the importance of prudent fiscal reforms and welfare reform, tradition compels Jews to speak out to ensure that these reforms are not made at the expense of helping the most needy. In all these ways, the Jewish Way stands in opposition to the consumerist values that drive free-market capitalism.

THE CHRISTIAN WAY

Traditional Christian interpretations of the Tanak and the New Testament present a plurality of understandings of human beings and society quite different from the understanding offered by free-market capitalism. Rather than an autonomous individual, the Christian Way presents individual human beings as embedded in a series of interdependent relationships— with God, with other persons, and with the rest of creation. This means that Christian conceptions of freedom—with the exception of Calvinist thought—differ profoundly from capitalist economic theory. Rather than freedom *from* the other, Christian understanding of freedom is primarily freedom *for* the other: for God, for other persons, and for the rest of creation. Dietrich Bonhoeffer captured the Christian understanding of freedom superbly when he wrote: "Freedom is a relationship between two persons. Being 'free-for-the-other,' because I am bound to the other. Only by being in relation with the other am I free."[14]

In other words "personhood" is a communal construction created by the ceaselessly changing interrelationships between human beings structured by the ideals of compassion and justice for all. So persons cannot be understood apart from community, or community apart from the individual persons of which it is composed; this relation between community and individuals thereby avoids the extremes of individualism—persons in competitive separation—on the one hand, and collectivism. It is within this relational context that Gen 1:28 is best understood: "God blessed them [Adam and Eve] and said to them, 'Be fruitful and multiply, and fill the earth and subdue it; and have domination over the fish of the sea and the birds of the air and over every living thing that moves upon the earth.'" As progressive Christians today interpret this verse, humans are called to be *stewards*: to take the gifts that have been entrusted to them by God and, through work, to cultivate these gifts so that they become more productive while remaining sustainable.

The creation account in Genesis also suggests that while God has provided enough to meet human needs, the creation needs to be cultivated so that human needs can be met continuously and sustainably. This requires making decisions about where to allocate resources in order to meet human needs while sustaining the earth God has created. This profoundly contrasts with the consumerist principle that "no person ever has enough." At their best, progressive Christian visions of economic life reject the assumption that human beings are constituted by wants that can never fully be satisfied. This is so because most interpretations of the Christian Way embrace the teaching that human fulfillment is found in following the Way of the historical Jesus confessed to be the Christ of faith.

Accordingly, individual human beings must never be reduced to that economic fallacy of misplaced concreteness economists refer to as "the consumer." Of course human beings must "consume" to remain alive. So too must all sentient beings. This is just a fact of evolution. But "to consume" does not define human nature or the nature of other life forms with which human beings share planet Earth. Moreover, given that both the Jewish and

14. Bonhoeffer, *Creation and Fall*, 23.

Christian Ways affirm that God has created and sustains a world in which there are enough resources to meet everyone's needs, poverty is a result of human institutions, economic and political policies, human greed, and environmentally unjust behavior.

Both the Jewish and Christian Ways affirm the importance of place: in the creation narrative, humans are placed within the *cosmos*; Israel is given a *specific* land as a gift from Yahweh; and Jesus comes to a particular people at a particular time, in a particular place. In contrast, capitalist theory pays little attention to the role of place in economic life, and neoclassical economics has reinforced this tendency. In this connection, Shuman notes, "For traditional economists and their critics, place was beside the point. The basic unit of analysis for microeconomics was the firm, and for macroeconomics the nation. Community, a level of organization somewhere in between, didn't really fit."[15] That is to say, the vast majority of economists understand "place" neutrally, as something like a blank canvas or a checkerboard upon which pieces can be moved with ease according to whatever economic goals are in mind. For economists "place" means "labor and capital," and these concentrations of labor and capital should be moved to those places (those locations) where labor and capital can achieve the highest rate of return (i.e., the highest salary, wage, or profit).

The relational orientation found in the opening creation narratives is maintained throughout the Tanak and the New Testament—from Israel's vocation to be a light to the nations to Jesus's summary of the law to the final pages in Revelation where the vision of a new heaven and a new earth and a new Jerusalem is one of *shalom* ("wholeness," "peace"). What is portrayed in these texts is a vision of the relational harmony between God, humanity, and the rest of creation. Although it is God who will ultimately bring this relational vision to its consummation, even in the presence of sin, human beings are called to partner with God in working towards this end.

In biblical texts that deal explicitly with economics these interdependent relationships are clear. For example, the Jubilee laws (Lev 25) put limits on the unequal distribution of land, which would have been the principal form of productive capital in ancient Israel. Economic institutions are to support the relational orientation of life in general, while economic life is subordinated to a relational understanding of life; when, for example, the historical Jesus taught the interdependent principles of love of God and

---

15. Shuman, *Going Local*, 291–92.

love neighbor, he was pointing to the priority of persons-in-community over individual financial wealth.

The essential difference between capitalist economic theory and the Christian Way in all its pluralism can be summarized as follows: Current economic theory, and in particular free-market capitalist theory, assumes a highly abstract and distorted view of personhood. Individual human beings are "autonomous beings," with the result that community is portrayed as a collection of atomized individuals rather than as communities embedded in network of relationships. The irony is that despite placing the individual at the center of its schema, capitalism actually downplays the *personal* in economic life.

THE ISLAMIC WAY

Of the 350 "legal verses" in the Qur'an, known collectively as *ayat al-ah-kam*, 140 relate to doctrine and devotional practices, including the duties involved in the Five Pillars: declaring there is no god but God and that Mohammed is God's messenger, which prefaces most Muslim prayers; offering five times daily the ritual prayers known as *salāt*; giving alms, or *nammaz*; fasting (*imsak*) on holy days, particularly during Ramadan; and making a pilgrimage to Mecca once in one's lifetime (*hājj*). Another seventy verses are devoted to marriage, divorce, paternity, child custody, inheritance, and bequests. Rules about transactions such as sales, loans, usury, rents, and mortgages are the focus of another seventy verses. Roughly thirty verses are devoted to justice, equality, evidence, citizens' rights and duties, governmental obligations to citizens, and around ten verses concern strictly economic matters.

The Islamic Way began its history in direct confrontation with hard economic, and at times military, confrontation with the urban centers of the Byzantine Empire and the ancient cities of Persia, now known as Iran. The other religious Ways I have described in this book all began their historical evolution within agricultural environments and later in their evolution confronted the economic structures of large cities that were part of large, powerful empires and so adjusted their founding traditions to this urban reality accordingly.

But Islam originated in direct confrontation with urban economic and political power structures from the very beginning of its history. Mecca and Medina were centers of an important caravan trade linking both towns to the great cities of the Middle East, particularly Constantinople. Mohammed's tribe, the Qur'ish, took kickbacks from all the caravans that passed through Mecca, and Mohammed married a wealthy widow named Khadijah, fifteen years his senior and his first convert to Islam, who owned and operated her own caravan trading business. There isn't much agriculture going on in a desert—unless you live along the Nile River or the Tigris and Euphrates Rivers—but trading in goods and services allowed the Arab people a means of making a living. Thus the economic rules and duties of the Qur'an reflect desert life in contact with city life.

One economic policy of the Prophet was a ban on charging fees and rents and a ban on permanent buildings in the market of Medina, where only tents were allowed for the purpose helping poor traders. Social responsibility in commerce was stressed in early Islam so that business practices could not include usury. Interest rates were not allowed, nor were investors permitted to escape the consequences of failed ventures. All financing was equity financing, or *musharaka*, so that borrowers did not have to bear all the risks and costs of failure. The goal of *musharaka* is not allowing economic practices to break communal harmony. Muslims, according to *shari'a*, are also prohibited from trading in forbidden goods such as wine, pork, and gambling.

Accordingly, "Islamic economics" refers to rules of economic transactions that conform to *shari'a* decisions or *fiqh* consistent with the Qur'an and recorded in the Sunna. Traditional Islamic economic jurisprudence focuses on what is required, encouraged, discouraged, or permissible in cases not related to the Qur'an or the Sunna. Islamic jurisprudence is always tested and measured by what Muslims believe is the revealed word of Allah in the Qur'an and the practice of Mohammed as portrayed in the Sunna. This standard applied to economic issues like property, money, employment, taxation, and essentially all things economic.

By the mid-twentieth century, campaigns promoting specifically Islamic economics spread throughout the Middle East as a reaction to Western political and economic imperialism. The central features of Islamic economics can be summarized in three principles: (1) the foundations of economic goals and economic practice must be derived from the Qur'an and the Sunna, (2) *zakat* (alms) and other taxes are collected, and (3) the prohibition of interest (*riba*) is observed.

Advocates of Islamic economics describe it as neither capitalist nor socialist but as a middle way between these two economic theories, an ideal economics with none of the drawbacks of these two Western systems. Among the important claims made for an Islamic economic system is that the gap between the rich and poor is reduced as prosperity increases for all: hoarding wealth is discouraged, the wealthy are taxed through *zakat*, lenders are encouraged to take risks in profit sharing and venture capitalism, hoarding food for speculation is discouraged, and confiscating land unlawfully is forbidden.

Detractors often criticize Islamic economics as an "invented tradition" that is too unrealistic given the realities of the global market and current economic structures. But between the eighth and the twelfth centuries, many quite advanced concepts developed in Islamic cultures—concepts of production, investment, finance, economic development, and property use known as *hawala*. In addition, Muslims invented an informal value-transfer system; a system of trusts known as *waf*; contractual systems relied upon by merchants; a widely circulated common currency, a system of banks, promissory notes, and bills of exchange (*mufawada*); and agricultural technologies. Islamic concepts of money, property, taxation, and charity included *zaqat* (taxing monetary and agricultural produce for the purpose of allocating these resources to aid the needy), *gharar* ("the avoidance of uncertainty" in contracts), and *riba* (the prohibition of interest on loans).

These concepts, like others in Islamic jurisprudence, evolved from the communal memory of prescriptions, anecdotes, examples, and the words of the Prophet, all gathered together and systematized by various commentators. Sometimes other sources such as *al-urf* (the "custom" of the Prophet Mohammed), *al-'aql* ("reason") or *al-ijma* ("consensus" of the *'ulma* or "jurists") were employed. In addition, as Muslims encountered Western systems of law and economic practice, Islamic jurisprudence imported areas of law that roughly correspond to Western laws about contracts and torts. A self-described "Islamic economics" "emerged" in the 1940s, and as of 2004 Islamic banks have been established on these principles in over seventy countries, and interest has been banned in three: Pakistan, Iran, and Sudan.

THE HINDU WAY[16]

From ancient times the Hindu Way concluded that economic needs and spiritual practices are interdependent. Thus the creation of wealth is of vital importance in the creation of community. One of the four goals of Hindu

16. See Vinot, *Handbook of Hindu Economics and Business*, ch. 1.

life is *artha* (something like "material well-being"). The duty (*dharma*) of every householder is to support one's family and one's local community. The "householder stage of life" is considered extremely important because it provides financial support for the maintenance of the whole community. Accordingly, the Hindu Way generally encourages the creation and accumulation of wealth so long as wealth is then redistributed for the well-being of one's community, the exception to this principle being those who have renounced the world and have taken up the way of asceticism (an optional "third stage of life").

Thus the Hindu Way generally upholds economic principles and practices as a means for the upkeep of the family. All individuals in a community have an obligation to help the poor, and not doing so reaps negative karmic affects for both the poor and the wealthy. Accordingly, most forms of the Hindu Way advise affluent devotees not to hoard wealth but to act as the stewards and distributors of wealth. And on a more contemporary international scale, the Hindu Way encourages rich nations to extend an immediate helping hand to Third World countries, not merely because the world is now a global village, but because the whole world suffers as a result of their poverty.

In two main contemporary forms of the Hindu Way, there exist two contrasting understandings of charity. Dualistic Hinduism (Dvaita Vedanta) is a monotheistic form of Hinduism that teaches that individual souls are separate from God and recycle in samsaric existence according to their deeds until they achieve final rebirth in Paradise. Practicing compassion and performing acts of charity (*daya*) are the central means of pleasing God. The nondualistic tradition of the Hindu Way (Advaita Vedanta) teaches individual selves (*ātman*) and Brahman are metaphysically identical (The Great Identity or Brahman equals Atman). Accordingly, the real reason we should help others is that essentially we are one in our nonduality. It is Brahman alone that is self-incarnated in things and events at every moment of space-time, so helping others is in reality helping ourselves. The right attitude is to look on the poor as Brahman, and it is our duty (*dharma*) to serve Brahman in the here and now through our charity

THE BUDDHIST WAY

The term "Buddhist economics" was coined by E. F. Schumacher in 1955 during his work as an economic consultant to then Prime Minister U Nu of Burma.[17] According to Schumacher and other Buddhists with whom I have spoken about economics, five aspects of the Buddhist Way distinguish it from the Cartesian worldview assumed by capitalist economic theories. First is capitalism's assertion of human self-interest as the driving force of economic activity. Buddhist teachings regarding "nonself" (*anatta* in Pali and *anātman* in Sanskrit) are a direct challenge to the central role self-interest plays in capitalist economics, particularly in free-market capitalism. Buddhist teachings, as confirmed by the experience of meditation, conclude that all things perceived by one's senses are not actually "I" or "mine." Accordingly, because there exists no permanent "self" in "self-interest," capitalist economic theory is an illusion. As a result, capitalism is the primary cause of communal suffering (*duḥkha*), which is brought about by clinging (*taṇhā*) to the idea of permanent economic self-fulfillment of individual selves that do not exist.

A second difference between capitalism and the Buddhist Way is that capitalist economic theories center on maximizing profits for individuals and corporations, whereas Buddhist economics stresses minimizing suffering for all sentient and nonsentient beings. Consequently, the Buddhist Way stresses compassionate sensitivity to the universal suffering that all sentient beings experience simply by being alive, as we undertake socially engaged action directed toward reducing universal suffering. All economic activity should be directed to this end, which exemplifies two items of the Noble Eightfold Path: Right Conduct and Right Livelihood.

A third difference between capitalist economic theories and Buddhist economics centers on desire. Capitalist economic theories focus on fulfilling human desires for wealth by encouraging more and more consumption of the goods and services that corporations create and sell. In Buddhist teaching, every form of desire entails clinging to the fiction of permanent selfhood. Accordingly, the desire for wealth, while increasing the illusory sense of permanent well-being for those who are rich, only increases the suffering of the vast majority of human beings as well as the life forms with which all human beings share planet Earth.

A fourth point of difference relates to the market. While capitalist theories advocate maximizing markets to the point of saturation, Buddhist economics aims at minimizing violence. Capitalist economic theories ignore future generations and the environmental structures of the natural world because their "votes" are not important in terms of purchasing

17. See Schumacher, *Small Is Beautiful*.

power. Nonwealthy individuals, particularly the poor kept in poverty by the market, along with the environment supporting all life are not represented in current capitalist systems because preference is given to the wealthiest stakeholders. Therefore, the market is not an unbiased economic force that truly represents the economy.

Buddhist economists also advocate "nonviolence" (*ahimsā*). Community-supported agriculture is one example of nonviolent economics. Buddhists believe such enterprises foster trust, help build value-based communities, and bring people closer to the land. Achieving sustainability nonviolently requires restructuring the dominating configurations of contemporary business. It also entails deemphasizing profit maximization as the ultimate economic motive while emphasizing small-scale, locally adaptable, substantive economic practices.

A fifth difference between Buddhist economics and capitalist economics is that capitalist economic theories and practices focus on maximizing instrumental use where the value of any entity is determined by its marginal contribution to the production output and economic gains of corporate stockholders. Buddhist economists affirm that the real value of persons-in-community is neither realized nor given importance by capitalist theory and practice. Accordingly, economic practices should reject instrumentalism and refocus on creating caring businesses that are economically rewarded by the degree of trust given by persons-in-community.

The sixth and seventh differences separating Buddhist economics and capitalist economic theory and practice are (6) that capitalism assumes that bigger is better while Buddhist economics assumes small is better and less is actually more; and (7) capitalist economics centers most of its energy on gross national product while Buddhist economics focuses on communal well-being.

From the point of view of a Buddhist economist, the most rational way of life is being self-sufficient and producing local resources for local needs. Dependence on imports and exports is economically justifiable only on a small scale. Economic development should also be independent of foreign aid and deemphasize the use of nonrenewable natural resources, which should be used only when most needed and even then with utmost care. In short, Buddhist economics asserts that capitalist economic development inflicts violence on the natural environment that sustains the well-being of all sentient beings and therefore is contrary to the Buddhist Way.

THE CHINESE WAY

Even though capitalism was not part of the experience of the ancient writers of the religious Ways I have summarized, central ideas in these traditions offer important insights that contemporary persons need to apprehend when making economic decisions for the common good. In the case of the Chinese Way, two points are particularly important: the Confucian stress on human interdependence and our interdependence with nature, and the Daoist stress on finding balance between *yin* and *yang* and intentionally living in balance with the Dao's natural rhythms.

From a Confucian perspective, human beings should always give primary attention to fulfilling duties to family and close social relations before all else, including economic security. As the *Analects* has it, "A man of humanity, wishing to establish his own character, also establishes the character of others."[18] That is, according to Confucius, it is not a matter of consumption but a need for the basic elements of life that makes life "alive." The Confucian Way is a way of caring for others in community. The model for building such community is the family.

More concretely, the Confucian Way begins with every individual's obligation to relieve the distress of those closest to him or her. If a relative is in dire straits, family members should offer help. If a neighbor is having trouble, those nearby should offer assistance. Those who have more than they need to provide the basics for their own family should contribute their resources to the wider needs of their community. While other forms of charity are important, the Confucian Way emphasizes close personal connections.

As for government economic policy, the Confucian Way focused less on "economic stimulus" and more on assistance in hard economic times, as the following passage from *The Mencius* illustrates:

> Treat with respect the elders . . . of [the] family, and then extend that tenderness to other families. Treat with tenderness the young of [your] own family, and then extend that tenderness to other families . . . Let mulberry trees be planted about the homesteads with their five *mou*,[19] and the men of fifty will be able to be clothed in silk. Let there be timely care for fowls, pigs, dogs, and swine, and men of seventy will be able to have meat to eat. Let there be no neglect in the timely cultivation of the farm and its

18. *Analects* 6:28, in Chan, *Sourcebook in Chinese Philosophy*, 31.

19. About one-sixth of an acre.

hundred *mou*, and the family of eight will suffer no hunger. Let serious attention be paid to education in school, elucidating the principles of filial piety and brotherly respect, and the gray-haired men [i.e., the poor] will not carry burdens on the roads.[20]

So as a religious Way emerging within the context of an agriculturally based society, the Confucian ideal suggests that the state under the leadership of emperors should follow "Heaven's Mandate" by establishing economic policies that facilitate the acquisition of the agricultural necessities supporting human community, of which the family is the model for understanding what "community" means. And according to Mencius, any emperor failing in this part of "Heaven's Mandate" loses the right to rule, at which point the common people have the right to revolt in order to support any person who can demonstrate his credentials for assuming "Heaven's Mandate" and establish a new dynasty. In other words, all economic decisions of government, from the emperor to local officials at the village level, should be primarily concerned with the communal acquisition of the goods and services that support just and compassionate community.

Again, reflecting the economics of an agriculturally based social system, the Daoist Way is less interventionist than the classical Confucian Way. The concept of "acceptance" has a specific meaning in the Daoist Way, which can be illustrated by Zhuangzi, also widely known in the West as Chuang Tzu. In the "seven inner chapters" of *The Zhuangzi*, which according to most scholars may have been either written by Zhuangzi or preserved as oral tradition by his first- and second-generation disciples and later set in writing, acceptance is one of the central themes of his Daoist teachings. In all matters, including economic matters, acceptance should be practiced at all levels of human interaction in one's relations with nature or with society. If and when individuals and governments are beset by economic hardship, this situation should not be interpreted as some kind of political sign for revolution. Those beset by economic hardship should not read into their situation any sort of moral lesson. No individual and no government are "bad" because of economic difficulties. Both are merely caught up in the turbulent transformation of that aspect of the Dao contemporary human beings termed the "economy." Consider the following:

> What is acceptable we call acceptable, what is unacceptable we call unacceptable. A road may be made by people walking on it; things are so because they are called so. What makes them so?

20. "The Mencius," in Chan, *Sourcebook in Chinese Philosophy*, 61.

> Making them so makes them so. What makes them not so? Making them not so makes them so . . . For this reason, whether you point to a little stalk or a great pillar, . . . the Way makes them all into one. Their dividedness is their completeness; their completeness is their impairment. No thing is either complete or impaired, but all are made into one again. So he [the Sage] has no use for [categories], but regulates all to the constant. The constant is the useful; the useful is the passable; the passable is the successful; and with success, all is accomplished.[21]

In other words, just let economic circumstances wash over you, discover what is possible, and let go of what is not possible at that moment by flowing with the Dao's back-and-forth movement between *yin* and *yang*. Never cling to the past, and allow the economic realities of the present to flow as you ride Way. But a Confucian critique might be, flowing with the Dao doesn't help pay the rent, and capitalist economics would certainly agree. The Daoist Way has little room for economic instrumentalism.

Obviously, the classical Confucian Way is not the mystical way of classical Daoist tradition and practice. The Confucian Way is a way of social engagement with the hard realities of political and economic necessities required in constructing human community. Yet both the Confucian Way and the Daoist Way are founded on the same worldview, but interpreted differently. And the Chinese people for twenty-five-hundred years have united both Ways into "the Chinese Way."

As a way of inner experience and withdrawal from the issues of statecraft and economic concerns in order to follow the ebbs and flows of nature, the Chinese people typically understood the Daoist Way as a *yin* way of life. But as a Way of social engagement with the hard political and economic realities of communal life, the Confucian Way is a *yang* way of life. The Chinese people seemed to have "listened" to the worldview underlying both Ways and tried to live at the ever-changing point of balance between the Confucian and Daoist Ways. For there are times when following the Confucian path of active social engagement is the best course for creating just, compassionate community. But there are also times when flowing with the course of natural events is the wiser, more just and compassionate communal path. One needs to be always sensitive to the Dao's energy flows whether one identifies with the Confucian Way or the Daoist Way. Seeking

---

21. Watson, trans., *Complete Works of Chuang Tzu*, 40–41.

to not live at the extremes of *yin* and *yang*, the Chinese Way counsels an exquisite balancing act between both extremes.

CONCLUSIONS IN PROCESS

To conclude that the religious Ways I have summarized offer identical understandings of the meanings of "compassion," justice," and "community" would not only be worse than ahistorically wrong, but would also undercut all dialogue between these religious Ways. Nevertheless, each religious Way affirms in its own uniqueness that compassionate justice is the glue that transforms human beings from a mere collection of persons into community, even as the structure of existence such community actually entails differs from religious Way to religious Way. Confucian community and Daoist community are not identical—neither are Jewish, Christian, Islamic, Hindu, or Buddhist communal structures identical. Yet each religious Way struggles to create compassionately just communal structures of existence. Each affirms that individual human beings are persons-in-community.

This is one of the amazing facts of the religious pluralism that currently characterizes planet Earth. According to the process-theological perspective that underlies my particular participation in the Evangelical Lutheran Church of America, the fact of this reality requires all persons, including secular persons, to engage in interreligious dialogue as a process of passing over and returning to their particular traditions creatively transformed.[22] In such a dialogue, one passes over into a religious Way other than one's own, learns what needs to be learned, and returns to one's own particular religious Way creatively transformed. An important focus of such dialogue is how each Way engages in issues of individual and social justice in the struggle to create compassionate communal structures of human existence.

Such an interreligious dialogue is absolutely necessary as a collective means of socially engaging the ideology of capitalism, particularly free-market capitalism. The economic ideology supporting the corporate economic exploitation of human community everywhere is for the sole purpose of consuming this planet's natural resources for the 1 percent of wealthy corporate stockholders, as the rest—the 99 percent—are treated as isolated individuals consuming the natural resources upon which all life depends. In other words, free-market capitalism encourages the death of this planet in order to satisfy corporate and consumerist greed.

22. See Ingram, *Passing Over and Returning*.

While this conclusion in process does *not* entail abandoning all forms of capitalism, *it is* an argument for revising capitalist economic theory by abandoning the centrality of isolated individuals in separation from other individuals engaged in an economic survival of the fittest. This economic model of the fallacy of misplaced concreteness must be replaced by the more realistic model of persons-in-community working for the common good of all. Corporations and the academic discipline of economics itself will most certainly resist abandoning contemporary economic dogmas and unsustainable practices. Abandoning current capitalist practices will require resources from all of humanity's religious Ways that focus on the pluralist meanings of justice, compassion, and community.

5

Interreligious Dialogue and Resistance against Ecological Injustice

Human beings have been struggling to create just, compassionate communities at least since shaman-artists painted the figures of animals on the walls of caves in Lascaux, France, some twenty to thirty thousand years ago. In all times and in all places against all odds this struggle continues. It is humanity's religious Ways that everywhere took the lead in this struggle, more often than not against conventionally religious human beings and institutions that populate all of the world's religious Ways. So the call to live justly and compassionately in community has come from a minority of powerful voices within humanity's religious Ways, often in opposition to their own tradition's domination systems.

Examples abound: the entire Mosaic tradition celebrated by Jews as interpreted through the oracles of the eighth-, seventh-, and sixth-century-BCE prophets; the historical Jesus whom Christians confess as the Christ of faith; Mohammed; Gautama the Buddha; the Daoist sages; and Confucius. Contemporary examples abound as well: they include every rabbi I've ever met, Martin Luther King Jr. and Thomas Merton, Badsha Khan and Mahatma Gandhi, Wing-tsit Chan, the prophetic voices inhabiting all of humanity's religious Ways, and the wise lay followers of these Ways struggling throughout their lives to live justly and compassionately.

Consequently, human resources for resisting the economic and political systems responsible for global environmental injustice are clearly available in humanity's religious Ways, but only if some means can be found to unite faithful persons inhabiting these Ways in a more unified struggle for the common good that includes all sentient beings. Of course, the meanings of "compassion," "justice," and "community" are tweaked differently in different religious Ways. Yet we have at all times and in all places been "told what is good": be compassionate, do justice, and build community between human beings and between human beings and nature. But, how are we to do this? is the perennial question. The structures of just, compassionate communities are always evolving in relation to historical times and places. Like all things caught in the field of space-time, no specific communal structure is permanent even if founded on compassionate justice. Accordingly, this chapter's thesis is that in an age of religious pluralism that is simultaneously an age of horrific environmental danger to human life, the question, how are we to do this? needs to be the focal point of interreligious dialogue between the world's religious Ways. But it may be environmentally too late, and this dialogue needs to begin now.

Of course, important interreligious voices and organizations are engaged in dialogical confrontation with environmental issues. Again, examples abound: the Dalai Lama; Elizabeth A. Eaton, the presiding bishop of my Lutheran community, the Evangelical Lutheran Church in America (ELCA); the Lutheran World Federation; the United Methodist Church; the Presbyterian Church (USA); Justin Welby, Archbishop of Canterbury; the bishops of the American Episcopal Church; the membership of the Society for Buddhist–Christian Studies; the Forum on Religion and Ecology at Yale University, founded by Mary Evelyn Tucker and her husband, John Gimm; the Buddhist Churches in America; the Religious Action Center of Reform Judaism; the Center of for Theology and the Natural Sciences in Berkeley, California; the Center for Process Studies in Claremont, California; and online organizations such as Pando Populus (http://pandopopulus.org)— and the list could go on across the planet.

But among the most prophetic voices of resistance against the economic and political causes of climate change is that of the current pontiff of the Roman Catholic Church, Pope Francis. His third encyclical, *Laudato Si* ("Praise Be to You"), calls upon all Roman Catholics, Christians in other faith communities, faithful participants in non-Christian communities, secularists, and natural scientists to engage in a global dialogue on climate change and on the basis of this dialogue to take immediate and strong action to address and redress it causes.[1] John Cobb describes the center of

---

1. The title of this encyclical comes from a canticle of Saint Francis of Assisi, "*Laudato*

this dialogue as a call for "an integral ecology of environment, economy, and equality" because Pope Francis "asks for nothing short of a complete transformation in how we teach, how we govern, how we do business, how we think, and whom we include."[2]

Francis of Assisi, Pope Francis's namesake, referred to the earth as "Mother Earth" and as a "Sister" who sustains and governs all life inhabiting the earth. But, as Pope Francis writes:

> The sister now cries out to us because of the harm we have inflicted on her by irresponsible use and abuse of the goods which God has endowed her. We have come to see ourselves as her lords and masters, entitled to plunder her at will. The violence present in our hearts, wounded by sin, is also reflected in the symptoms of sickness evident in the soil, in the water, in the air and in all forms of life. This is why the earth herself, burdened and laid waste, is among the most abandoned and maltreated of the poor; she groans in "travail" (Rom 8:22). We have forgotten that we ourselves are dust of the earth (cf. Gen 2:47); our very bodies are made up of her elements, we breathe her air, and we receive life and refreshment from her waters.[3]

What Pope Francis is referring to in this section of *Laudato Si* is climate change caused by global warming. He calls for an "integral ecology of environment, economy, and equality." And although Pope Francis's theological worldview is grounded in two thousand years of Roman Catholic theological reflection and practice, he calls for a worldwide ecumenical dialogue—what I have referred to as socially engaged interreligious dialogue—about how human beings are shaping planet Earth that includes Protestant and Orthodox voices and non-Christian voices along with secularist voices in conversation with natural scientists. He writes:

> We need a conversation which includes everyone, since the environmental challenges we are undergoing, and its human roots, concern and affect us all. The worldwide ecological movement has already made considerable progress and led to the establishment of numerous organizations committed to raising awareness of these challenges. Regrettably, many efforts to seek concrete solutions to the environmental crisis have proved ineffective, not only because

Si, mi Sigore"—"Praise to You My Lord." See Francis's *Canticle of the Creatures,"* in *Francis of Assisi: Early Documents*, I, 113–14.

2. Cobb, "Preface," iv–v.

3. Pope Francis, *Praise to You*, 9.

of powerful opposition but also because of a more general lack of interest. Obstructionist attitudes, even on the part of believers, can range from denial of the problem to indifference, nonchalant resignation or blind confidence in technical solutions . . . All of us can cooperate as instruments of God for the care of creation, each according to his or her own culture, experience, involvements, and talents.[4]

For Pope Francis, this "ecumenical dialogue" needs to focus of the human roots of the ecological crisis while developing an "integral ecology" founded on "the principle of the common good" for human life and the life forms with which we share this planet, in direct confrontation with the economic and political domination systems that harm the environment as they reduce millions of human beings to lives of poverty.[5]

Accordingly, a good part of *Laudato Si* is devoted to calling us to a global socially engaged interreligious dialogue that (1) insists that climate change caused by global warming is real and caused by human beings; (2) calls for the rapid conversion of global economies from coal, oil, and gas to renewable forms of energy such as solar and wind power; (3) insists that the first victims of global warming are the poor; and (4) keeps in mind that given the extent of the climate and environmental damage human beings have already caused we no longer have room for a slight shift from fossil fuels. We simply have to make an immediate leap to renewable power because the life of all sentient beings is at stake.

The scientific facts about global warming have been confirmed and are now well known. According to climate scientist James Hanson's "Timeline for Irreversible Climate Change," the primary cause of global warming is "overreliance on fossil-fuels."[6] Fossil fuel reserves are at present declining, yet large oil corporations continue tearing the earth apart to extract every last drop of oil and lump of coal. In their search, oil companies like Exxon/Mobil, Shell, and British Petroleum proudly advertise that they are drilling the depths of the oceans and exploring the most pristine environments to feed consumerist addiction to fossil fuels while increasing the profits of their stockholders. But this corporate greed is "as insane as a drug addict trudging through an arctic blizzard in search of a fix."[7]

---

4. Ibid., 18–19; see also ch. 5.

5. Ibid., chs. 3–6.

6. Hansen, "Timeline for Irreversible Climate Change."

7. Ibid.

If humanity is to continue surviving on a planet similar to the one on which human civilization evolved thousands of years ago and to which all present life on Earth is biologically adapted, the ratio of carbon and oxygen (CO_2) must be reduced from its present 385 ppm (parts per million) to at most 350 ppm. Hansen estimates that peak CO_2 can be kept to about 445 ppm, with large estimates for remaining oil and gas reserves, if coal is phased out by 2030, although some "peakists" believe that the planet is already at peak oil production because nearly half of the globe's readily extracted oil has already been consumed. Consequently, Hansen argues that we must pursue a radically different direction premised on the possibility that achieving 350 ppm CO_2 or lower in this century will minimize the possibility of passing tipping points that spiral out of control: continued global warming caused by the emission of greenhouse gases into the atmosphere, the complete disintegration of ice sheets and the rapid rise of sea levels, the acidification of oceans and the loss of coral reefs, and the mass extinction of countless species. The list goes on: rising temperatures coupled with an increase in dangerous weather-related events like hurricanes and abnormally heavy rainfall and flooding, interrupted by increasing periods of heat waves; rising sea levels flooding coastal cities and farming communities.

Along with the destruction of coastal cities, wetlands and rainforests will be severely damaged, and communities depending on these lands for their subsistence will be forced to migrate and in the process will clash with settled inland communities for whatever resources are left to sustain life. Poverty will increase as resources that sustain life become scarce so that the poor are forced into violent confrontation with wealthier human beings just to stay alive. And the tragedy is, all this planetary destruction is directly linked to current economic practices that sell the practice of consumerism like an addictive drug.

So the future is grim indeed, and we had better act now to undertake the necessary steps to reverse global warming. Such collective action must begin by breaking our consumerist addiction to fossil fuels, because a large fraction of CO_2 emissions will linger in the atmosphere for many centuries. In other words, "we're not going to get back the planet we used to have, the one on which our civilization developed."[8]

But what might the structure of an ecumenically interreligious dialogue look like? And how might such a dialogue help human beings resist the consumerist causes of global warming? These are important

8. McKibben, *Eaarth*, 16.

and controversial questions. What follows are my particular suggestions regarding the structure of interreligious dialogue and how interreligious dialogue might become a resource for resisting the corporate causes of global warming. Hoping to energize the dialogue, in this chapter I add my voice to an interreligious discussion of how the religious Ways of the world together might become a force to be reckoned with against consumerist economic and political systems pushing life on this planet closer and closer to extinction.

That interreligious dialogue has emerged in the last twenty years as a primary topic of interest among faithful religious persons and scholars is beyond question. For me, a convenient beginning point for reflection on the structure of dialogue is the process that John S. Dunne described as "passing over" and "returning," which means the structure of dialogue is "Socratic."[9] By this I mean that authentic dialogue demands a twofold realization of those participating that, first, they are ignorant of the "truth" so that, second, they desire to undertake the risky search for this "truth," which is, as Socrates believed, the first step toward "wisdom." Awareness of ignorance and the beginning of wisdom merge as we engage in dialogical encounter. Once we experience this "push," dialogue entails two polar movements: (1) "passing over" into the faith and experience of persons dwelling in religious Ways other than our own, and (2) "returning" to the "home" of our own religious Way, now deepened and creatively transformed by the "odyssey" of the encounter.

Of course, much depends upon the worldview and commitments from which the odyssey begins. Normally, Christians pass over and return to a particular faith tradition within the Christian Way, a Muslim to a particular tradition of the Islamic Way, or a Buddhist to a particular school of the Buddhist Way, such as Zen or one of the many schools of the Tibetan or Chinese traditions. It is this form of dialogue that I have referred to as a "theology of religions" model.[10]

However, numerous religious persons actively involved in interreligious dialogue find themselves unable to return to the home of their original religious Ways. They are deeply influenced, perhaps overwhelmed, by the reality of religious pluralism and the relativizing of religious faith and practice this reality occasions. Such persons include scholars as well as faithful individuals trying to make sense of the realities of postmodern

9. Dunne, *Way of All the Earth*, preface, ch. 1.

10. Ingram, *Wrestling with the Ox*, 15; see also Cornille, *Criteria of Discernment*.

existence. They may wear the labels of a particular religious Way, or they may be committed secularists. Some find themselves wearing dual religious identities—for example, finding themselves, like Paul F. Knitter, both Roman Catholic Christian and Zen Buddhist.[11] They are deeply influenced, perhaps overwhelmed, by the realities of contemporary religious pluralism. Such persons, including me, end up as theological pluralists.

But the general consensus of those involved in interreligious dialogue is that the goal is renewal and creative transformation of one's own particular religious Way. Otherwise, why bother with dialogue at all? Accordingly, four conditions must be met before conversation with persons whose faith is different from our own can be called dialogue. First, all ulterior motives must be excluded. Approaching another person's religious standpoint with ulterior motives is a monologue, not a dialogue. For example, comparatively studying Buddhist faith and practices with the intention of demonstrating the superiority of the Christian Way undermines the integrity of both the Christian Way and the Buddhist Way. Interreligious dialogue is not missionology.

Second, interreligious dialogue requires being engaged by the faith and practices of one's dialogical partner. Nothing important emerges from interreligious dialogue unless our own perspectives are challenged, tested, and stretched by the faith and practices of our dialogical partner. Approaching persons as advocates of our own religious Way transforms dialogue into a monologue. Any truth we seek is not found because we have absolutized our particular religious Way as the ultimate truth before the dialogue begins. It is just this sort of religious imperialism that blocks mutual creative transformation.

But third, interreligious dialogue demands clear, critical, and empathetic comprehension of one's own religious Way. A good part of the openness of engagement with others requires being engaged by the truth claims of one's own religious Way. Without a point of view, a perspective grounding one's life, dialogue can only be a formless sharing of ideas in which we merely state what we believe, as persons inhabiting other faith traditions reciprocate in kind. In the end, nothing important or valuable is achieved. To use a musical metaphor, we need to hear the music of our own faith tradition before we can hear the music of religious Ways other than our own.

Finally, the truth sought through interreligious dialogue is relational and centered on the goal of creative transformation. "Truth"—whatever

11. See Knitter, *Without Buddha*.

religious, theological, philosophical, scientific, political, or economic label it wears—can have no institutional or confessional boundaries in a universe that the natural sciences are now portraying as governed by universal relativity and probability. This surely implies that Lutheran Christians like myself can share their faith without presupposing the inferiority or falsehood of their, for example, Buddhist or Muslim brothers or sisters. Interreligious dialogue must grow out of our common humanity as persons whose sense of what it means to be human expresses itself through different yet real encounters with a Sacred Reality named differently by different religious Ways. "Truth" is falsified into doctrinal systems of half-truths when we approach one another through abstract labels like "Christian," "Buddhist," or "Muslim." Put another way, theology must not become ideology. Or in an analogy appropriated from Zen Buddhist tradition, doctrines, beliefs, or worldviews are like fingers pointing at the moon. We should never confuse the pointer with the reality to which it points.

Because religious human beings engage in interreligious dialogue for different reasons, it is useful to summarize three interdependent forms of dialogue that have been evolving over the past thirty years, particularly within the context of Buddhist–Christian dialogue:[12] conceptual dialogue, socially engaged dialogue, and interior dialogue. Most Protestant and many Catholic theologians, for example, focus on conceptual dialogue in their encounter with non-Christian religious Ways; a growing number now emphasize socially engaged dialogue, and Roman Catholics and a growing number of Protestants interested in such disciplines as contemplative prayer and non-Christian meditative traditions focus on interior dialogue.

But it must be kept in mind that these three types of dialogue are utterly interdependent. Conceptual dialogue requires understanding the specific doctrines and teachings as well as of the practice traditions of one's dialogical partner that motivate his or her social activism. Christians walking in social protest against injustices like global warming (practicing socially engaged dialogue) must understand what motivates non-Christians engaged in social action—their doctrinal commitments and spiritual disciplines.

The focus of conceptual dialogue is doctrinal, theological, and philosophical. It concerns a religious tradition's self-understanding and worldview. In conceptual dialogue, for example, Jews, Christians, Muslims, Buddhists, Hindus, Doists, and Confucianists compare theological and philosophical views on such questions as the nature of ultimate reality,

12. See Ingram, *Buddhist–Christian Dialogue*, 10–15.

human nature, and suffering and evil. Partners in conceptual dialogue might compare the nature of the historical Jesus confessed to be the Christ of faith to the historical Buddha as a teacher of the way to achieve Awakening; or dialogue partners might compare Mohammed's role as Allah's final prophet and the structures of existence according to the teachings of Confucian and Daoist sages.

Since issues of social, environmental, economic, and gender justice are not religion specific, socially engaged interreligious dialogue is now emerging as a major theme of interreligious dialogue worldwide. Faithful religious persons in all of humanity's religious Ways have apprehended common experiences of injustice as well as common resources for working together to liberate human beings and nature from the economic and political forces of systemic oppression.

The term "social engagement" first emerged within the context of Buddhist social activism. The Vietnamese Buddhist monk Thích Nhất Hạnh coined the term "social engagement" in the 1960s as a description of the Buddhist antiwar movement in his country. Many progressive Christians picked up this term as a description of their social-activist traditions. In the case of the Buddhist Way, the heart of social engagement is the practice of nonviolence that shares much in common with the social activism of Mahatma Gandhi and Martin Luther King Jr.

Part of the meaning of Awakening according to Buddhist teaching is experiential awareness of "dependent co-arising" (*prutīyā-samupt-pāda*), which in turn engenders "compassion" (*karunā*) for all sentient beings. That is, compassion is awareness that in a mutually interdependent universe, the suffering of others is the suffering of all. Compassion in turn engenders nonviolent social engagement, the goal of which is to relieve sentient beings from suffering everywhere at all tines and in all places. Here, although Buddhists do not share the theistic worldview that supports the prophetic narratives in the Tanak, the meanings of the prophetic concepts of justice and compassion are quite similar to concepts within the Buddhist Way. It is also worth noting that the calls for justice and compassion in the literature of the Israelite and Judahite prophets and in the Qur'an are not too dissimilar to the conception of justice in the Buddhist Way, despite the fact that the Buddhist Way and the Ways of Jewish, Christian, and Islamic faith engender different structures of existence.

In the work of most Christians engaged in dialogue with non-Christian Ways, conceptual dialogue and socially engaged dialogue often engender interior dialogue. Interior dialogue concentrates on participating in meditative and spiritual disciplines other than one's own and on reflecting upon the resulting experiences. As a Lutheran Christian, I have been privileged to receive instruction in several Buddhist traditions of meditation and to have participated in daily prayer rituals in several mosques; I have also taken part in purification rituals in Shinto shrines and have received instruction in the Spiritual Exercises of Ignatius Loyola from Jesuits and in the Jesus prayer from several Benedictine nuns. Such practices have led me to appreciate more deeply what Luther was railing against in his criticism of the Catholic Church of his day, but more importantly to the realization that clinging to doctrines transforms theology into an ideological shell game having no relevance for engaging the social, political, and environmental forces now in process of destroying this planet and its life forms. We, all sentient beings, are in this together and we had better bring our collective selves to the struggle "before its too late."[13]

But before socially engaged interreligious dialogue can evolve into something more than a mere conversation on religious doctrines and teachings between persons of different faith traditions—about which most people on this planet do not give a damn—it must take on a prophetic character in its focus on environmental issues such as global warming and its causes—blame for which lies mostly with large corporations whose only purpose is to increase human consumption. Accordingly, the natural sciences must be brought into this dialogue as a "third partner."[14] This is so because since the seventeenth century, a cosmology has been evolving about the physical processes at play in the universe—from subatomic quantum forces to planets to galaxies to the universe itself to all living things. All religious traditions, of course, have their own cosmological worldviews. But these must not be abandoned: rather they must be adjusted to what the scientific community continuously discovers about the physical forces at play throughout the universe in general and on planet Earth in particular. The physical processes involved with global warming and other environmental problems must be understood so that appropriate corrective actions might counter and perhaps reverse global warming. As far as I can tell, a prophetic socially engaged interreligious dialogue with the natural sciences might be the best hope human beings have for addressing ways of curtailing the global consumerism fueling global warming.

13. Cobb, *Is It Too Late?*.

14. Ingram, *Buddhist–Christian Dialogue*, ch. 1

Yet it is clearly a fact that millions of human beings across this planet admire capitalist economic systems whose very existence depends upon convincing people to consume the earth's natural resources. This admiration ignores the fact that capitalist economic theories leave out of consideration the needs of human beings, the destruction of the environment, and the destruction of nonhuman sentient beings. The engine driving capitalism is consumerism. Consumerism is the religious faith of most people on this planet, including the vast majority of persons now participating in the religious Ways of humanity.

Consequently, before resistance can be effective in changing the current economic paradigm, abstract disciplines of inquiry must be transformed from specialist disciplines concerned with narrowly defined areas of human knowledge into interdisciplinary enterprises concerned with the wide spectrum of what human beings and nonhuman beings actually experience. This will require nothing less than abandoning contemporary overreliance on the ways that areas of human knowledge are balkanized into specialized departments in contemporary universities and colleges. Interreligious dialogue is by nature an interdisciplinary practice to its very core: it involves theologians, philosophers, textual scholars, pastors, priests, nuns, imams, rabbis, and ordinary folk sitting in pews or on meditation mats or prayer rugs—all talking with one another from their particular perspectives about the fundamental issues facing the planet: war, environmental destruction, poverty, homelessness, and, yes, global warming.

This last paragraph reflects my agreement with economist Herman E. Daly and process theologian John Cobb.[15] Two points must be understood. First, economics, like theology or my field, history of religions, is very successful. Second, like other social scientists, economists choose to copy the primarily deductive methods of the natural sciences. So like those trained in natural sciences, economists are trained to abstract a discrete area of human experience from the totality of human experience, treating its abstractions as having no connection with the remainder of what is really going on in the world and therefore having no economic importance. On this assumption, economists have invented their own highly abstract methods for studying their subject matter. This means that economists abstract from everything other than that to which a monetary value can be assigned. This means that economics as practiced since Adam Smith's day

15. See Daly and Cobb, *For the Common Good*, ch. 6.

is deeply involved in what Alfred North Whitehead called "the fallacy of misplaced concreteness."

Here's why. The problem with disciplinary approaches to knowledge is overreliance on the centrality of *method*. Of course every disciplinary method entails abstractions, and the current prominence of mathematization in economics serves only to deepen its particular abstractions. Of course, there is nothing inherently wrong is using methods; in fact, all disciplined thinking is methodological. But there is no such thing as a single right method. And to the extent that a discipline, say economics, reduces its subject matter to mathematical models, to that extent it places limitations on itself for confronting new and unexpected realities—"boundary questions" about which it is either blind or that it simply ignores.[16]

The compartmentalization of knowledge into academic disciplines and subdisciplines began during the European Enlightenment that presupposed the mind–body dualism of René Descartes (1596–1650). But the hegemony of academic disciplines in the United States is quite recent. Prior to World War II, higher education was not dominated by academic disciplines, as it was in European universities. Instead, liberal arts colleges, many associated with the various churches that populate Christian tradition, encouraged interdisciplinary thinking. This was so because the education going on in these colleges was centered on areas of thought that defined human beings and culture as "human." In other words, the "liberal arts" included a number of academic disciplines taught as an integrated collection of methodologies in order to expose students to different ways of thinking about the meaning of being human.

But after the Second World War, the disciplinary organization of knowledge took off so that today, economists, scientists, historians, philosophers, and professional theologians are trained to think, speak, and act as if their particular disciplines cover the whole range of what is to be known. So successful has this project been that now contemporary universities and colleges are places where students are sent to major in an academic discipline for the purpose of obtaining entry into a high-paying position after graduation. Meanwhile, interest in interdisciplinary education is declining along with interest in the liberal arts: music, literature, art, philosophy, history, religious studies—what Confucian tradition calls "the arts of peace." Furthermore, the imposition of methodological boundaries on knowledge continues unabated in contemporary higher education, as more and more

---

16. See Smith, *Scandalous Knowledge*.

universities adopt corporate organizational models for structuring their curriculums.

The problems of disciplinary abstractions increase because of the centrality of method in a particular discipline's self-understanding. Every method requires abstractions, and the current preeminence of mathematization in the social sciences, particularly in economics, is a good illustration. As previously noted, there is nothing inherently wrong with using methods. All disciplined reflection on anything is methodological, and critical thinking requires a plurality of methods. In fact, self-consciousness about method is something all good teachers try to teach their students. But here's the hiccup: no single method exists for thinking about anything. This is so because to whatever degree a method redefines its subject matter in terms of what yields to its method(s), to that degree abstractions are reinforced and made invisible. "If the only tool one has looks like a hammer, then everything begins to look like a nail."[17]

In point of fact, there are many methods, that is, rational "ways" of understanding most anything. Some of these are designed to open us up to the full, concrete, interdependent structures of existence, as do Buddhist practices of mediation. Others are designed to open us to the full, nondual presence of God, as Christian centering prayer and contemplative prayer do—both of which bear much congruence with what Buddhists refer to as mediation. But whether one is engaged in an academic method or a meditative method, we encounter through these methods whatever we are trained to expect to find. Trained in the methods of economics, economists find what their methods train them to expect to find. Everything else is ignored as simply irrelevant to economic reflection. Likewise, Buddhists and Christians interpret the meaning of their meditative experiences according to the traditions that train them.

So here's the point. Knowledge comes to us through the disciplines in which we are trained to know anything. But we are always constrained by our methods everywhere at all times and in all places so that knowledge is always partial and incomplete. There is always more to learn and understand, whether one is a physicist, an economist, a Christian theologian, or a Zen master. We are limited by our methods of understanding anything, which is why "boundary questions" keep popping up like popcorn in all fields of inquiry. Yet knowledge is not achieved without methods of

17. Daly and Cobb, *For the Common Good*, 128.

knowing even if the knowledge obtained is always partial and incomplete. There is always more to know.

But what does all this have to do with interreligious dialogue as a means of resisting the economic-political systems that are the root causes of climate change in general and global warming in particular? The simplest answer to this question is that the practice of interreligious dialogue is not constrained by any single method that defines its goals or subject matter. Rather, the practice of interreligious dialogue requires a plurality of methodologies, each working in concert.

Practitioners of interreligious dialogue include theologians, philosophers, historians of religions, social scientists, economists, political scientists, psychologists, and should also include scholars in the natural sciences. In other words, interreligious dialogue needs the whole range of human religious experience as interdependent with other forms of experience: political, economic, medical, social, economic, psychological—and the list could go on. Dealing with boundary questions, as I have noted, requires interdisciplinary ways of thought and practice. The existentially most important boundary questions facing human beings today are environmental issues related to global warming: rising CO_2 and methane levels in the atmosphere, rising sea levels due to the melting of ice caps in the Arctic and Antarctic, drought, forest fires, deforestation, animal and plant extinctions, migration of species from original habitats, weather patterns, and on top of everything else, water pollution. Then there are the worldwide social, political, and economic consequences involved in reversing, if possible, the environmental damage we have already foisted on planet Earth.

No one has written more clearly about these issues than Bill McKibben. He notes that not only does global warming damage people's bodies, but also their minds. He writes:

> Doctors in countries with bad heat waves report an upswing in psychosis. One young man was so bothered by the link between warming and drought that he convinced himself millions would die if he drank a glass of water. In the wake of Hurricane Katrina, the increase of severe mental illness doubled in the affected areas, with 11 percent of the population suffering from post-traumatic stress disorder, depression, and a variety of phobias.[18]
>
> Then there is the reality of global poverty. One of the most difficult obstacles to actually doing anything about global warming is global poverty.

18. McKibben, *Eaarth*, 75.

Just as we come into this crisis with an infrastructure deficit and an overhang of debt, so we also suffer from a justice deficit that will slow any attempt at action. In 2008, after a careful study of prices for goods and services in developing countries, the World Bank issued a new set of numbers: 1.4 billion people, it found, lived well below the poverty line than previously estimated. What defines the poverty line? $1.25 a day . . . And the remedy? Exactly the things in shorter supply: India needs to increase farm productivity, all developing countries must spend more on education, and Africa requires stability above all, to encourage investment. Instead, when the market for biofuels exploded in 2008, we saw food riots in thirty-seven countries and as many as 100 million Africans moving back into poverty.[19]

As I am writing this chapter in 2016, these statistics have increased and are guaranteed to continue increasing unless and until governments and the large corporations of the developed countries on this planet, who are most responsible for global warming, work together through the United Nations to confront these environmental disasters.

So what positive influence can socially engaged interreligious dialogue between the world's religious Ways that includes the natural sciences as a third partner have in stopping global warming, or at least in slowing it down and mitigating its effects? First, all religious Ways, as specified by the teachings of each, call us to create just, compassionate communities that seek the common good for all persons and the natural environment supporting all life. The goal of interreligious dialogue is mutual creative transformation interdependent with the creative transformation of the natural environment. This is the primary obligation all religious persons share without exception.

But entering interreligious dialogue as a means of asserting the superiority of our particular religious Way over others fosters a parochial theological monologue that transforms our faith perspectives into ideological prophylactics meant to protect us from whatever it is we reject. Too many religious prophylactics are at large in the world. My own tradition, a Lutheran version of the Christian Way, has been especially guilty of confusing theological reflection with ideology. Theological exclusivism, like Christian fundamentalism and fundamentalisms ingredient in all religious Ways, are simply irrelevant when it comes to resolving environmental issues such as global warming.

19. Ibid., 76.

Second, a socially engaged interreligious dialogue by its very nature focuses on the creation of just, compassionate communities working for the common good founded on the interdependence of all things and events. All the religious Ways I have described in this book assert that the foundation of justice, compassion, community, and the common good rests on the interdependence of all things and events since "the beginning." Violence, injustice, wars pitting human communities against each other, the oppression of women and the poor, and the exploitation of the earth's natural resources by the corporate and politically greedy tears this interdependence apart. Here rests the source of violence, injustice, and the destruction of human community and our community with the environment. All one has to do to confirm the truth of this observation is read the daily newspapers, catch the daily news on television, or bury one's attention in social media. All of humanity's religious Ways have told all human beings in their own peculiar ways "what is good." It is time that we listen and act accordingly.

Socially engaged interreligious dialogue begins with awareness that justice issues involving the creation of compassionate community and environmental issues such as global warming are utterly interdependent. For example, environmental injustice caused by corporate greed, Western development programs, and the exportation of large corporations from Europe and the United States to so-called underdeveloped countries have resulted in the depletion of agricultural resources in these countries. This has forced farmers off their farms into cities to find jobs at very low wages and to live in cramped corporate slum housing after hours of working in dangerous factories, with very little medical care as they make cheap products for European and American consumption. As more and more working-class jobs are sent abroad, workers in the United States and Europe experience poverty. And the cycle continues.

Linked to the corporate exploitation of the poor is the depletion of the planet's natural resources, particularly water resources. For example, in Africa there is little access to drinkable water, a situation worsened by the growing pattern of privatizing water resources and turning water into a commodity subject to the laws of the marketplace. Related to water shortage is a general loss of biodiversity because of the loss of woodlands and forests along with the accompanying loss of many species needed for food and medicine. The problem rears its head in the growing extinction of not only large mammals and birds but also many microorganisms necessary for the life of ecosystems. This is a particularly difficult problem for the two major rain forests on this planet: the Amazon and Congo rain forests. In addition, the great aquifers and the planet's glaciers and coral reefs, which shelter millions of life forms, are also disappearing. As aquifers, glaciers, and reefs disappear, other water sources are being polluted, and agricultural and fishing communities are being destroyed. Water pollution affects entire communities politically as well as entire countries.

Third, there are millions of religious people on this planet whose collective resistance, protest, and activism just might be the force necessary to compel the leaders of economic and political institutions to cease working for the good of the wealthy and powerful and instead to work for the common good of all human and nonhuman beings on planet Earth. Since socially engaged interreligious dialogue does not require doctrinal agreement between religious human beings, the focus on compassionate, just community for the common good emerges as a force unifying the diversity of religious human beings with a unity inclusive of persons who identify themselves as secularists or atheists. For when it comes to environmental issues such as global warming, water depletion, or deforestation, we are all in this together. And doctrinal and philosophical differences fade away like snakes shedding skin as politicians and their corporate allies are forced to pay attention and change the way economic systems work.

Of course, faithful religious people are motivated to confront the issues of environmental and social injustice by their own particular faith traditions. For example, as a Lutheran Christian, I affirm God as creator of the universe—although I think God continually creates through the processes of evolution—and that human beings, collective images of God, are to be stewards of God's creation. In other words, we must preserve natural resources, not consume them until the earth is no longer capable of supporting life. My Jewish and Muslim friends would largely agree with this even as they would nuance my theological propositions differently. My Buddhist friends, who are nontheists, are motivated by the doctrines of nondual compassion for all life forms as interdependent realities. Some of my Hindu friends believe that all things and events are incarnated expressions of one Sacred Reality named Brahman. While in conceptual dialogue with such people I have been challenged, theologically stretched, and deepened in my own faith through the process of passing over and returning. But in confrontation with the political and economic causes of injustice, doctrinal, philosophical, and theological differences become utterly irrelevant. Let me cite one example.

During July and August of 2015 hundreds of protesters surrounded a huge Royal Dutch Shell drilling platform scheduled to be towed from Portland, Oregon, to the Gulf of Alaska to begin exploratory oil drilling. Protesters surrounded the rig in hundreds of kayaks in an attempt to stop an icebreaker from towing it up the Willamette River to the Columbia River to the Pacific Ocean and north to Alaska. Some protesters hung from a bridge crossing the Willamette River from seats attached to ropes long enough to block the icebreaker. Protesters hung from these seats in shifts for three days until they were arrested. I can't really argue that the protest worked, because the ocean-going tug pulling the rig eventually arrived in the Gulf of Alaska. But on September 28, 2015, Royal Dutch Shell announced it was abandoning its quest to become the first company to produce oil in Alaska's arctic waters, because it "failed to find enough oil to make drilling profitable."

None of the protesters in Portland worried about the theological commitments or philosophical worldviews of the persons carrying signs, paddling around an oil rig in kayaks, or hanging from the Willamette Bridge. Yet the protesters included Christians from all denominations, Jews, Muslims, and Buddhists from different traditions, as well as people who defined themselves as "spiritual but not religious," along with committed secularists, agnostics, and atheists. And a source for global warming, melting ice sheets, and the exploitation of Native Americans was stopped, or at least slowed down. This is a small achievement, perhaps, but one that could be amplified worldwide if the majority of religious human beings populating planet Earth simply rose up and said, no! to the current economic paradigm and the corporate interests fueling consumerism.

And this is the point: Interreligious dialogue in all its forms, but particularly socially engaged dialogue, is premised on respect for and understanding of the plurality of religious and cultural riches of different communities of people—their worldviews, their art and poetry, their music, their interior life and spirituality. If we are truly committed to the development of an ecological awareness capable of stopping the damage we are now doing to the environment, every discipline of the natural sciences must be included as a third partner in the dialogue. No way of knowing or form of wisdom can be left out, which of course includes the religious Ways of humanity.

Finally, the goal of a worldwide interreligious dialogue that includes the natural sciences as a third partner is the creation of sustainable communities founded on justice and compassion. Since "communities" are pluralist social structures, no particular communal structure of existence can be dominant. Each of humanity's religious Ways seeks just, compassionate community, but in its own distinctive way. But the common search for just, compassionate communal structures constitutes a rich diversity of human interrelationships and human interdependence with the exquisite diversity of nonhuman life forms gracing planet Earth. For humanity's common good, this diversity of human relationships in a plurality of communal structures must be allowed to evolve into a deep awareness of the unity within our diversity. Only then will we fully understand that we are all in this together. Here lies the goal of interreligious dialogue. Achieving this goal for our and the earth's common good will not be easy because it requires active resistance against the present economic and political domination systems responsible for global warming as they collectively drive the consumption of this planet to its death. What such resistance might look like is the topic of the following chapter.

6

The *Praxis* of Interreligious Dialogue

So far I have been arguing that socially engaged interreligious dialogue focusing on climate change could become a powerful force of resistance against the human causes of the approaching disasters now threatening planet Earth. However, before becoming more specific about how such a dialogue might contribute to curtailing global warming, I must summarize the operating assumptions behind my thesis.

I can't remember reading any mainline Christian theologian who claims that God is interested in an abstraction called "religion." In fact according to the New Testament, God doesn't give a damn about "religion" but cares very much about the welfare of all sentient beings inhabiting planet Earth. So say the Tanak and the Qur'an, although Islam may be the only religious tradition that defines itself as "religion." In Arabic the word meaning something like "religion" is *din*, the practice of which is *'islām*, "surrendering" to God's will that we live justly and compassionately with each other and sustainably with all sentient beings. So it seems that the one and only test of the truth of a religious idea or practice is pragmatic; or as the historical Jesus is reported to have said, "You know them by their works." A truthful religious idea, doctrine, practice, or religious experience leads directly to practical acts of compassionate justice. As the collective religious Ways discussed in this book teach us, "compassion" is knowing by experience the utter interdependence of all things caught up in the field of space-time so that the suffering of any sentient being is one's own suffering. Compassion engenders justice expressed as active social engagement with the world in (if possible) nonviolent struggle against any and all systemic social, economic, political, and religious domination structures fostering injustice

In other words, if one's experience and understanding of the Sacred makes one kinder, more empathetic, more impelled to act justly through concrete, nonviolent acts of justice and lovingkindness—at least as far possible in a universe in which life must eat life to survive—then one's understanding of the Sacred corresponds to reality, to the way things really are. I think this is true whether one is a Christian, a Jew, a Muslim, a Hindu, a follower of the Daoist or Confucian Way or both, or a participant in a nontheistic tradition such as Buddhism, or an avowed secular humanist, or an atheist. But if one's notion of the Sacred has made one unkind, brittle of spirit, belligerent, cruel, or self-righteous, or if one's notion of the Sacred has led one to kill in the Sacred's name, then one has a false understanding of the Sacred. I also think this is true whether one is a Christian, a Jew, a Muslim, a follower of the Daoist or Confucian Way or both, or a Hindu, or a participant in a nontheistic religious tradition such as Buddhism, or a secular humanist, or an atheist.

Raimundo Panikkar lists a number of problems encountered by anyone engaged in theological reflection about God.[1] I can't speak for other persons, but my own theological reflections correlate well with this list:

1. We cannot speak of God meaningfully without first having experienced an interior silence.

2. Speaking or writing about God is unlike anything else we can speak or write about, because God is not a "thing" or an "object" that can be captured by the doctrines and practices of any religious Way.

3. Reflection about God, written or spoken, is a discourse about our entire being in interdependence with all beings in the universe and with God.

4. Discourse about God is not the monopoly of any single religious Way because God is not the monopoly of any particular religious Way. All forms of religious imperialism are illusory.

5. Meaningful theological reflection is always reflection by grace through faith alone. Of course as a Lutheran, I take this as given.

6. Theological reflection is not reducible to doctrines believed to be literal descriptions of what God is or is not, because theological language

1. Panikkar, "Nine Ways not to Talk about God."

is always symbolic. This is the reason why fundamentalism in all of its expressions is misguided; fundamentalist ideas say more about the persons who hold them than about God.

7. Thinking and speaking about God is a pluralistic discourse that cannot be limited to a particular religious tradition.

8. We cannot understand or signify what the word "God" means in terms of a single perspective, which is why interreligious dialogue is the heart of authentic theological reflection.

9. All theological discourse completes itself by entering what Thomas Merton called "the Silence." The religious Ways that exist today all agree on this point in their own distinctive ways. This is the "interior space" where you shut up in order to be opened up by the mystery that is God that theological reflection invariably conceals.

Panikkar was writing about the transcendence of God. But the problem is that a God that is completely transcendent cannot be conceived or spoken about and would be utterly superfluous to the universe in general and human experience in particular. A completely transcendent god is a denial of divine immanence. So while God is ultimately ineffable mystery, this does not mean we cannot say anything about God even though we cannot say everything there is to say. Just because we cannot know or say everything does not mean we cannot know or say something that is meaningful. Transcendence and immanence for God and for beings created in the image of God are interdependent. Or restated in the language of process theology, "transcendence" is God's "primordial nature," as Whitehead described it, while God's "immanence" refers to God's "consequent nature" that is always active for the common good within the particulars of space-time since the beginning of space-time.

So human beings have been instructed in what the common good requires by the teachings and practices of humanity's religious Ways, by an incredible plurality of teachings and practices. From this plurality emerges a profound thirst for compassionate, just communal structures working for the common good. Given this worldwide search, interreligious dialogue must focus on (1) the common meanings of compassion, justice, and community within each religious Way; (2) the different nuances of meanings of "compassion," "justice," and "community"; (3) the pluralist meanings of "the common good"; and (4) how to create political, social, and economic structures of just, compassionate communities working for the common good reflective of humanity's unity in diversity.

It is important, therefore, that socially engaged interreligious dialogue be grounded in conceptual and interior dialogue. Understanding the teachings of religious Ways other than our own and participating in their distinctive practices—prayer, meditation, communal activities like feast days and celebrations of important events—deepens our understanding of our own faith. Such dialogue may even require some to leave the home of their original faith communities and enter the faith communities of their dialogical partners. The practice of interreligious dialogue is not for the fainthearted. Whether we continue along our own religious Way or begin moving along another, we become less parochial as we become open to religious diversity.

Of course there is a good deal of religious nonsense in all religious Ways. Fundamentalism is a dangerous distortion of the Christian Way. Radical Islamic groups are not groups that follow the Qur'an's explicit injunctions against terrorism and the oppression of women. Contemporary oppression of Palestinians by the government of Israel does not reflect the Torah's or the prophets' injunctions to compassionately and justly struggle for the common good of all. Violence, racism, war, the oppression of the poor by the rich, and the oppression of nature are not justifiable by any religious Way and are distortions of religious faith and practice whenever and wherever they occur. Interreligious dialogue exposes both the distortions of religious faith as well as the creatively transforming teachings and practices of humanity's religious Ways. So dialogue entails becoming "wise as serpents." Speaking for myself, the more I have known faithful persons living at the depths of their distinctive religious Ways, the more my own particular Lutheran faith has been deepened, stretched, and creatively transformed.

Perhaps two personal experiences will clarify what I have written in the preceding paragraphs. The first is an event that took place at Simpson College in Indianola, Iowa, where I was a member of the faculty from 1966 to 1975.[2] One evening in 1973, one of my Muslim students called to ask if he could come to my home for some advice about a research paper he was writing for my seminar on religious pluralism.

"Of course," I said to Abdul. As we sat in my living room discussing his thesis that Sufi mysticism, like Thomas Merton's "monastic dialogue," could

---

2. I originally described this event in Ingram, *Passing Over and Returning*, 127–28.

provide a "bridge" by which Muslims might be able engage in an interior dialogue with Christian mystical experience, Gail, my four-year-old daughter, entered the living room and sat on the couch next to Abdul. There were very few persons of color living in Indianola in those days. But she had seen persons of color before and was very curious about Abdul.

Abdul, who was from Egypt smiled, and said, "*Salam*, Little One."

Then Gail brushed her hand across Abdul's left cheek. "Daddy, she said, "why doesn't it come off?"

"Ask Abdul," I said.

When she did, Abdul relied, "Because this is the way God made me."

"Why did God do that?"

"Because God loves wondrous diversity."

So from an excellent undergraduate student and faithful Muslim my daughter received a lesson about the delusion of racism. As for me, it was Islamic confirmation of the justice of Martin Luther King Jr.'s leadership of the civil rights movement as well as a confirmation of my protest against an unjust war in Vietnam. It was also confirmation of the truth that we live in an interdependent universe where all human beings are brothers and sisters regardless of skin color or religious labels we choose or choose not to wear. All of us, therefore, as the Qur'an instructs, should "strive as in a race" for peace and justice "so that we may know each other" (Surah 49:13).

My second example comes from an experience at a conference in Japan on the *Lotus Sūtra*, hosted by Risshō Kōsei Kai, "The Society for the Establishment of Righteousness and Friendly Interchange."[3] The conference attendees were invited to a Sunday morning service at a local Risshō Kōsei Kai *kyōdan*, or "church." The congregation was seated in neat rows on *tatami* mats separated by an aisle, while we, as visitors, sat in chairs in the rear. The service began when the minister, dressed in a Methodist-looking black preacher's robe, entered as a line of young people processed two by two down the aisle singing hymns praising the *Lotus Sūtra*. (Many Japanese Buddhist lay groups picked up the Protestant flavor of this service from Protestant missionaries in the nineteenth century.)

Prior to his sermon, the minister invited a middle-aged woman to give her testimony. She tearfully recounted the conditions of her life prior to her

3. Risshō Kōsei Kai is one of the "New Religions" derived from the teachings of Nichiren's (1222–1282) interpretation of the *Lotus Sutra*. It was founded as a lay Buddhist movement in 1938 by Naganuma Myōkō (1889–1957) and Niwano Nikkyō (1906–1999) to spread the teachings of the *Lotus Sūtra*. I have recounted this event in Ingram, *Living without a Way*, 49–50.

conversion to Buddhism because of the influence RisshōKōsei Kai missionaries in her neighborhood: she told about the physical and emotional abuse
she sustained from her husband, her years of drug addiction, her rejection
by her children and relatives, and her life of poverty as a prostitute. But after
she converted to Buddhism, she said, her "negative karma was turned into
positive karma": her husband no longer abuses her, her children and family
now love her, and she no longer engages in prostitution to make financial
ends meet. In other words, this woman blamed herself for her own abuse
by men.

But then in a long sermon in Japanese, so did the minister. As I sat
listening to his sharp condemnation of the woman's life before she became
a member of Risshō Kōsei Kai, reconfirming the woman's self-blame for
her own abuse, I whispered to my friend, Mark Unno, "Am I hearing this
correctly?"

Mark, who is a Pure Land Buddhist and one of the finest scholars of
Buddhism in America or Japan, whispered, "Yes. Shut up!"

After the service ended, we were invited to meet the minister for tea
and pastries. Mark went directly to the minister and dressed him down
for using Buddhism in such a sexist manner to condemn a very troubled
woman who had experienced abuse from her husband and the other men
in her life. "She was not responsible for her abuse," he told the minister.

According to my worldview, while the testimony of this woman might
have been a story of her experience of creative transformation because of
her conversion to Buddhism, I also witnessed the power of creative transformation in the prophetic words and actions of a Buddhist scholar and
friend. Christian tradition has too often been a source of oppression, blam-
ing women for the abuse received at the hands of male clergy, husbands,
and laymen. Sadly, sexism is rampant in all of the world's religious Ways.

Of course the preceding two narratives reflect my own particular experiences. Therefore they cannot be taken as normative for anyone else.
They are merely anecdotes that might energize others to undertake their
own interreligious dialogues. But the individual experiences of thousands
of people engaged interreligious dialogue might constitute a powerful collective energy that might force the corporations and the politicians they
own to begin addressing the causes of climate change and to initiate economic and political policies that might halt the environmental disaster to
which the whole planet is heading. One voice, as Mahatma Gandhi and
Martin Luther King Jr. knew, is powerless to change the status quo. But the

combined religious voices of human beings taking part in socially engaged interreligious dialogue are not powerless. Two examples will serve to illustrate my point.

The first example is contemporary Christian conceptual dialogue with Buddhists, which has generated interest in the relevance of Buddhist thought and practice to issues of environmental, social, economic, and gender justice. Since these issues are systemic, global, interconnected, and interdependent, they are neither religious-tradition specific nor culturally dependent. All human beings at all times and in all places have experienced these forms of injustice. Accordingly, Christians and Buddhists have mutually apprehended common experiences and resources for working together to liberate human beings and nature from the global forces of systematic injustice.

The Vietnamese Zen Buddhist monk Thích Nhất Hạnh is generally given credit for coining the term "social engagement" in 1963 as a description of the Buddhist antiwar movement in Vietnam. But in fact, the Vietnamese "Buddhist Renewal Movement" first coined "social engagement" as *nham fian Phat Giao* in the 1930s.[4] But it was because of Thích Nhất Hạnh's leadership of the Buddhist antiwar movement in the 1960s that "social engagement" emerged as the most common term describing Buddhist social activism. Some progressive Christian liberation theologians have also appropriated this term as a designation of Christian social activism.

The heart of Buddhist social engagement is the doctrines of interdependence and nonviolence. Interdependence (*pratīya-samuptpāda* or "dependent co-arising") is the doctrine that all things and events at every moment of space-time are constituted by their interrelationships with all other things and events, so that no thing or event exists in separation. All things and events are mutually cocreated by this web of interrelationships. Since these relationships are always in a state of processive change, impermanence ("nonself" or *anātman*) is ingredient in the structure of existence itself.

Part of the meaning of Awakening is experiential awareness of dependent co-arising. This awareness engenders compassion for all sentient beings. Compassion originates from awareness that in a mutually interdependent universe, the suffering of other sentient beings is the suffering of all. This realization, in turn, energizes action to relieve all sentient beings

---

4. Rawlings-Way, "Religious Interbeing," 56.

from suffering. Compassionate nonviolence is the ethical heart of Buddhist social activism.

Socially engaged Buddhists are uncompromising in the practice of nonviolence. For progressive Christians this has raised questions about the place of justice. "Justice" is a central theological category in Christian tradition, as it is for Jewish and Islamic traditions. But the concept of "justice" has not played a congruent role in the Buddhist Way. Christian social activism gives priority to compassionate engagement with the hard realities of the world as the foundation for the creation of just communal structures struggling for the common good. So for Christians, the question is to what extent nonviolent compassion toward all sentient beings (including aggressors doing harm to whole communities of persons or to the environment) is itself an occasion for injustice.[5] While justice is not identical with revenge, Christian traditions of social justice demand that those who do harm not get away with it, which means that the establishment of justice may necessitate the use of violent means.[6]

So the practice of nonviolent compassion as the ethical norm of Buddhist social engagement has forced progressive Christians to reexamine the relationship between love, justice, and violence, and involvement in the world's struggle for justice has energized Buddhists to examine the relationship between the practice of nonviolent compassion and justice. In point of fact, one of the important questions Buddhists have been debating over the past few years is whether a distinctively Buddhist concept of justice is even possible.

Although a number of Christian theologians are in dialogue with Buddhist traditions of social activism, Paul F. Knitter is perhaps the best known. Drawing on Christian liberation theology, Knitter argues that a common context exists from which religious persons representing different traditions, in this case the Christian and Buddhist Ways, can enter into dialogue. Knitter identifies this common context with Christian liberation theology's "*preferential option for the poor and non-person*," meaning the obligation to

---

5. Cobb, *Beyond Dialogue*, chs. 4–5.

6. An important exception to this is the Quaker tradition of social activism, which does not include the use of violence as a means of resolving justice issues. In this, Quaker tradition is congruent with nonviolent Buddhist social engagement. Quaker practices of social activism, more than most mainline Christian Protestant or Catholic traditions, are modeled on the historical Jesus's injunction to "turn the other cheek" when confronted with an aggressor.

work for and with the needy of this world.[7] Furthermore, Knitter argues, apart from commitment to and identification with the poor and oppressed in the global struggle for justice, conceptual and interior dialogue between Christians and Buddhists remains an elitist enterprise with little relevance to the lives of oppressed persons or to resolving environmental issues like climate change.

And this is the point: In the midst of struggle against repressive domination structures, what religious people believe or practice within their own particular communities simply dissolves into the background as struggle with social, political, economic, religious, political, and environmental justices issues commences. Of course, engaging in conceptual and interior dialogues with Buddhist or Jewish or Muslim friends is absolutely necessary. No religious Way has a market on truth about the Sacred that Christians, Jews, and Muslims name God. There is much to learn from persons dwelling in religious Ways other than our own. Yet while these two forms of dialogue can deepen one's own particular faith and understanding, the grip of religious imperialism drops away from one's particular faith perspective while one is walking a picket line or hanging from a bridge in Portland, Oregon to stop a deep-water oil platform from sailing to the Arctic. At this point differences in doctrines and disciplines dissolve into our common humanity.

My second example is an event that happened in Tacoma, Washington. Around midnight some years ago, a group of men painted anti-Semitic graffiti on the outer walls of Temple Beth El. As they drove away in a green pickup truck, one of the vandals threw a railroad flare at the building. A passerby saw the fire and called the fire department, which put the fire out before it seriously damaged the temple. The next morning, the Associated Ministries of Pierce County, Washington, organized shifts of men and women to surround Temple Beth El twenty-four hours a day until the culprits were arrested. This group was composed of local members from the mainline Protestant churches, the Roman Catholic Church, the Tacoma Buddhist Temple, and the Islamic Center of Tacoma. These folks not only surrounded the temple with their bodies as a protective shield, but they also formed work details to clean up the graffiti and repair the burnt outer walls. This protective work continued for a week and ended when the culprits were identified, arrested, and brought to justice. It is important to note that

---

7. Knitter, "Towards a Liberation Theology of Religions," 185.

the culprits were not teenagers but full-grown men with connections to the American Nazi Party.

While informal conceptual and interior dialogical conversations happened among the people protecting and repairing Temple Beth El so that, as the Qur'an affirms, people might "know each other," the energizing force propelling this effort to protect a synagogue was the perception of a common humanity unifying religious differences in the struggle to create just, compassionate, inclusive community working for the common good. In situations requiring justice and compassion, doctrinal differences and practices become, in the words of the Indian Buddhist logician Nagarjuna, "secondary truths."[8] The defining "secondary truths" of particular religious doctrines and practices need to be understood through conceptual and interior dialogue so that "we may know each other." The greater truth behind the diversity of religious teachings and practices is how secondary truths reveal the common humanity, as Confucians point out, that defines us as human. We are one in our religious diversity, "nondual" as my Buddhist friends put it.

Given the realities of climate change now affecting the structures of human communities worldwide, a legal framework must be established to set clear restrictions on carbon emissions. Implementing such a system will involve confronting and reforming contemporary capitalist economic systems. Politicians with vested interests in passing legislation allowing corporations to ravish the earth for ever scarcer resources must be recalled, arrested, or both, for taking bribes from oil or coal companies. All human beings must be weaned from the unbridled consumerism now devouring this planet's ecosystems. A means must be found for redistributing corporate wealth more equitably so that the majority of human beings on this planet can be lifted from debilitating poverty. In all this, capitalism need not disappear as an economic philosophy but need only be redirected from the goal of gathering wealth for the top 5 percent to the goal of achieving the common good for all human beings so that just, compassionate communities will have the chance to evolve. I agree with Philip Clayton and Justin Heinzekehr: capitalist economics must be reformed through the incorporation of "organic socialist principles," particularly through the redistribution of wealth as a means of lifting 95 percent of the planet's human population from poverty.[9]

8. Streng, *Emptiness*, 161–66.

9. Clayton and Heinzekehr, *Organic Marxism*, 216–23.

Socially engaged interreligious dialogue with the natural sciences as a third partner is already energizing politicians and government leaders in some countries to combat the structural causes of global warming at a national level. A prime illustration is China, whose people suffer from some of the worst environmental problems on the planet: polluted air (especially in Beijing but also in all China's major cities and throughout its countryside), water pollution, land contamination from industrial waste, and mass migration as people looking for factory work move from small family farms into overcrowded cities.

Since the end of World War II, China's leaders have pushed industrialization with little concern for the environment. In order to "catch up with the West," they have followed Western European and American capitalist models, reinterpreting them in light of China's Communist ideology. To accomplish this goal, the Communist leaders initiated suppressive purges of China's traditional culture, particularly of the Confucian and Daoist Ways along with the Buddhist Way. Modern problems demanded modern solutions that meant, according to the government's interpretation of a Marxist philosophy, a complete rejection China's classical religious Ways. The sheer pace of China's modernization certainly transformed the country into a strong, independent nation free from Western colonial interference. But the ecological price paid for its policies has been high, and now government leaders are cautiously bringing the Daoist and Confucian Ways back into a nationwide discussion about how the current environmental dangers ravaging China might be combated.

It has been a creative dialogue contextualized by conversations with environmental scientists in China as well as in North America and western Europe. Daoist teachings about occupying the midpoint between *yin* and *yang*—the place of nature's creativity—individually and socially, and Confucian wisdom that each human being embodies a common humanity are gaining support from government officials. That is to say, the current Chinese political leadership has actually applied these ideas in national and local efforts to confront pressing projects. These include combating global warming, rising sea levels, species extinction, and water pollution; redesigning cities through urban plans that create open spaces and preserve what open spaces already exists; and eliminating corporate farming in favor of family farming near villages. Part of this dialogue has even involved Christian, Jewish, Islamic, Buddhist, and Hindu members of the Center for Process Studies, founded by John Cobb in Claremont, California. It is the Chinese government's dialogue with these religious Ways through the interpretive filter of process philosophy that is currently energizing China's environmental movement.[10]

Perhaps the most important interreligious dialogue between the world's religious Ways is that now taking place under the auspices of the World Parliament of Religions. The first meeting of the parliament happened in Chicago during the World's Fair in 1893. In 2009, the parliament convened in Melbourne, Australia. The Melbourne parliament focused on Aboriginal issues as well as issues of sustainability and climate change through the lenses of indigenous spiritualities. The council for the Parliament of World Religions suggested at the Melbourne meeting that environmental issues and religiously inspired violence are interconnected. Consequently, it is necessary that all persons of faith work together for global peace and justice by exploring the causes of religious conflict, by creating cross-cultural networks for addressing issues of religious violence, and by finding ways to work together internationally to confront environmental threats like climate change. The council urged that the voices of indigenous people be included in the conversations and the work.

Socially engaged dialogues between persons of faith in all of humanity's religious Ways are certainly a powerful resource for developing strategies for direct confrontation with those corporations most directly responsible for creating the consumerist desire that is literally eating this planet alive. In particular, the producers of fossil fuels such as coal, petroleum, and oil are driving global warming. Furthermore, they own most of the politicians making legal decisions affecting the well-being of the citizens of all countries. The issues are systemic: capitalist economic theory and practices need to be humanized. Socialist understandings about more just wealth redistribution need to be included.

Those of us addicted to the automobile will be forced to rely on smaller, more fuel-efficient vehicles or electric vehicles as the planet's gas and oil reserves run out. There will come a time in the near future when the planet's human population will need to rely on mass systems of transportation, which will necessitate the rebuilding of national infrastructures to accommodate the increasing reliance on trains and buses that will be

10. Two hundred scholars in Chinese religious traditions, philosophers, environmental scientists, and students from China attended an international conference organized by the Center for Process Studies, June 4–7, 2015. The Center for Process Studies has organized a number of international Whiteheadian conferences on environmental issues several times in China as well as in western and eastern Europe.

required to transport massive numbers of people with rapid efficiency. Human beings will need to lean how to live simply and efficiently in small self-sustaining communities producing most of what they need—food, shelter, schools, hospitals, police and fire departments, and places for communal gatherings—while relying on larger communities for obtaining what cannot be produced locally. The goal is sustainable living for the common good, which will require finding more efficient means for recycling a community's waste and taking from for earth only that which is necessary for the quality of human life.

All of the above will require changes that few wealthy CEOs and politicians—and ordinary people addicted to consumerism—will be willing to make unless they are forced to do so either by peaceful protest or by circumstances on an ever-warming and ever more damaged planet. Communal action must be undertaken now before its too late.

According to Pope Francis, "human ecology is inseparable from the notion of the common good."[11] The "common good" refers to the totality of communal relationships that allow social groups and their individual members to achieve meaningful fulfillment. In the present context of globalization, where injustices abound everywhere and growing numbers of human beings are deprived of the basic necessities of life and are even considered expendable according to capitalist economic practices, the common good is most directly a summons to solidarity firmly based on a preferential option for the poor and oppressed. This entails a redistribution of the world's goods more equitably, based on the recognition of the dignity of the poor. The fact that the vast majority of human beings are forced into abject poverty creates an ethical imperative to work for the common good.

Furthermore, the common good must be extended to future generations. As the current global economic crisis makes clear, working for the common good cannot exclude those who come after us. Sustainable development is interdependent with intergenerational solidarity. The point is this: when we reflect on the kind of world we are leaving to future generations, we apprehend everything differently. The planet is perceived as a gift freely received that we must share with others. The earth can no longer be viewed in a utilitarian way as a place in which economic efficiency and production are entirely geared to human consumption. Intergenerational solidarity is not an option, but a requirement of compassion, justice, and

---

11. Pope Francis, *Praise Be to You*, 108.

community. Planet Earth is on loan to each generation, and must be passed on to the next one with minimal wear and tear.

I remain cautiously hopeful that it is possible to begin addressing the human causes of climate change and perhaps even to undertake the necessary political, economic, and technological actions that might slow global warming down and perhaps even reverse the worst of it. But the fact is that the pace of consumption and environmental change has so stretched the earth's capacity to sustain life that environmental catastrophes are likely to increase both in frequency and destructive power. The effects of the present imbalances caused by human activity can only be reduced by decisive collective action here and now.

7

What Can Be Done?

"Community" is an overused word. Economists and politicians on the Left and Right use it the way fast-food restaurants use salted fat to cover up the lack of healthy ingredients in their food. Religious persons who use the word are often just as ambiguous regarding the meaning of "community." Are they Christian communities, as my own Lutheran congregation declares, that are "open and relational," or are they more like "members-only clubs," or are they something in between? In the prophetic literature of the Tanak, the meaning of "community" is inclusively clear. The meaning is clear too in the teachings of the historical Jesus whom Christians confess to be the Christ of faith. The meaning is clear as well in the Islamic Way, the Hindu Way, the Buddhist Way, and the Daoist and Confucian Ways. As opposed to social clubs and mere collections of human beings living in the same neighborhoods, "community" names an inclusively compassionate, just social-political structure that strives for the common good of all. We need to rescue this meaning of community so that "community" becomes the most clearly understood word in our thinking in this contemporary time of religious pluralism and dangerous climate change.

So we have been told what is good numerous times. But here's the problem in our day and time. Access to endless amounts of cheap energy for those of us wealthy and lucky enough to live in an industrialized country has made us rich in comparison to the majority of the poor who live in unindustrialized countries. Throughout the rhythms of economic booms and busts I have experienced as an American citizen, I have always found enough economic resources for meaningful existence. So while I might experience sadness when someone three houses from me has entered bankruptcy, I am never inconvenienced, mostly because I probably have not met him or her. Capitalist economics treat all human beings as independently separate from other persons, the "rugged individual" of an Enlightenment ideal going back to Descartes. Our present economic system is designed to work without input from our neighbors next door or down the block. Because of cheap oil, our food arrives from great distances with little inconvenience to ourselves other than shopping for it at the local grocery store. Credit cards and Internet connections make it possible to order much of what we need or desire, and to have it left anonymously on our doorsteps. In short, we've adapted to a neighborless existence and the almost near disappearance of community—except perhaps in some religious communities like the Quakers, Mennonites, or Amish—or my community at Pointe of Grace Lutheran Church in Mukilteo, Washington.

It is obvious that capitalist economic theory and practice is the engine driving the corporate greed for increasing the profits of stockholders, while consumerism is the fundamentalist faith of most individuals on this planet. The desire for more and more stuff runs amok even among those who identify themselves as Christian, Jewish, Muslim, Hindu, Buddhist, Daoist, or Confucian. But in point of fact, consumerism is contrary to all the religious Ways of humanity. Based on the assumption that human beings are individuals-in-separation, the interdependent relationships foundational to the creation of just, compassionate communities working for the common good have no chance to evolve. We have been told what is good throughout human history, but like human beings in the past, we in the present have failed to pay attention.

So what can be done to reverse the ecological damage we are causing to this planet, before it's to late to do so? It will not be easy given the facts of institutionalized corporate greed, corporate ownership of many politicians globally, and our extreme consumerist desire to own more and more stuff that we don't really need. What is fundamentally required, accordingly, is a conversion to understanding that human beings are persons-in-community and taking collective action accordingly. That is, our culture, with its collections of rugged individuals-in-separation-from-other rugged-individuals needs conversion to the ideal of persons-in-community. Our collective egoism, what Christian tradition refers to as sin, must be overcome. The religious Ways of humanity through socially engaged dialogue across the globe are best positioned to undertake this conversion process since they conceive human beings as persons-in-community and call human beings to form communities of compassionate justice for the common good.

But my argument admittedly has a problem. While it is historically quite true that the religious Ways of humanity have done much to encourage protection of the environment, none has been very effective over long periods of time. The Daoist Way, for example, failed to protect China's mountains from deforestation. Nor have Jewish, Christian, and Islamic notions of stewardship done much to prevent human-caused environmental damage to the earth's land, water, and atmosphere. The hard truth is that from the beginning, economic concerns have determined how human beings have interacted with nature. Nevertheless, it remains possible that given the seriousness of the environmental dangers now facing us, a worldwide socially engaged interreligious dialogue might raise our collective consciousness enough that current human behavior might be reformed and environmental disaster averted.

I need to make two preliminary points in this regard. First, corporate and political interests, not to mention our consumerist addictions, are so entrenched worldwide that success might be impossible in our increasingly secularized world. Second, the religious pluralism ingredient in humanity's religious Ways means that the specific forms of community-for-the-common-good will be pluralistic social structures. Just as no particular religious Way can claim a market on truth about the Sacred, so there can exist no single institutional form of community-for-the-common-good, but rather multiple forms existing in dialogical conversation with other forms globally. The following conclusions regarding what can be done should keep these two observations in mind.

According to the current economic paradigm, the increase of wealth is inherently good. Wealth is the measuring rod for determining the well-being of a society and individuals within it. But religious communities counterculturally teach that putting wealth first is contrary to the ideal of compassionate, just community seeking the common good. The dominant economic worldview now running amok in most parts of the world presupposes, as Adam Smith did, that human life is fully explainable by the force of self-interest that he called "the invisible hand."

Consequently, "being rational," means seeking one's own advantage in a survival-of-the-fittest world. Even human relationships (for example, marriages and friendships) are judged by what individuals-in-separation can gain for themselves. Religious communities, particularly religious communities engaged in interreligious dialogue, can help persons apprehend that we are more than ego-centered beings governed only by self-interest, although certainly self-interested actions are part of humanity's evolutionary DNA. But just as we are not reducible to our DNA, neither are we reducible to economic self-interest. Such reductionism is an example of what Alfred North Whitehead called "the fallacy of misplaced concreteness." Participation in the life of a religious community teaches us that human beings are much more that economic or political abstractions. We learn this through disciplines like prayer, meditation, liturgical practices, interreligious dialogue, and community outreach to the poor and disadvantaged, to those who for whatever reason didn't measure up to economically defined success—because capitalism has no room or concern in its worldview for the poor.

So what can religious communities engaged in a planetwide socially engaged dialogue accomplish through resistance to the current economic order and the consumerist engine driving this order? What follows is not an exhaustive list, but here are a few examples. There are already sporadic actions happening as I write this chapter. Many consumers throughout the world have joined together in working to begin new kinds of capitalism. These groups are religiously diverse. They include progressive Jews, Christians, Muslims, Hindus, Buddhists, Daoists and Confucians (or persons who embody both Ways), as well as progressive secularists indifferent to religion but nevertheless willing to work side by side with religious people. Numerous organizations are dedicated to informing consumers about how products are made, who produces them, and how they affect the environment in which we live. There are also millions of consumers favoring so-called green or environmentally sound products of all sorts and seizes over those whose production harms the environment. "Green products" are often more expensive, but nevertheless more and more people are willing to make the necessary economic sacrifices to "go green." In the United States consumer groups are pressuring the Food and Drug Administration to label ecofriendly products. Publicity about exploitative working conditions in factories is also beginning to influence buying habits worldwide.

As consumers become more aware of the environmental costs of refrigerating goods for transport over long distances, more and more are choosing to buy locally. In addition to consuming more locally, many people are learning that they can live quite comfortably without overconsumption. It's a matter of simplifying one's life through the discipline of consuming less by not identifying oneself with what one owns. Living more simply and locally, buying locally, and purchasing green products that cannot be produced locally from larger communal entities—cities, states, and foreign countries—greatly contributes to sustainable communal life. Living locally where one knows one's neighbors has the effect of collectively lessening human impact on the environment supporting the local community.

However, when all is said and done, local actions like "buying green" and living simply in local communities will not reverse the ecological dangers faced by this planet. What *must* be done in the space of just a few years is to break our collective addiction to fossil fuels. Human addiction to fossil fuels is the driver of global warming. No individual, no community, no nation-state can escape the terrible environmental consequences. Simply stated, a way must be found to return the earth's atmosphere to 350 parts per million carbon dioxide in order to stabilize the planet in its current state of disruption. This may not be possible because of the way the fossil fuel industry uses the atmosphere as its private waste disposal by pumping millions of tons of carbon waste into it yearly.

This must stop. The means of doing so involves replacing fossil fuels—oil, gas, and coal—with clean forms of energy such as wind power, solar power, and nuclear energy. We will need to greatly reduce our reliance on automobiles by creating public transportation systems capable of moving large numbers of people efficiently and cheaply. This will require rebuilding infrastructures and replacing gas-driven cars with battery-powered cars and perhaps solar-powered cars. As anyone who has driven a car on the freeways of Los Angeles on a hot day in July understands, smog deposited into the lower atmosphere is a major cause of respiratory diseases. Coal plants and refineries poison the atmosphere as well, while the petroleum and coal industries add to the poisoning of the earth and its atmosphere through fracking and strip mining.

However, none of the individual and communal acts of resistance suggested in the previous paragraphs will have much positive effect without structural changes in relations between the corporate world and the politicians controlling government bureaucracies. Corporate lust for producing an incredible variety of stuff to feed consumerist desire for "the good life" has emerged as the dominate force in industrialized countries. In our time, Adam Smith's theory of capitalism has been pushed to such extremes that those elected to political office in democratic societies are literally owned by the few billionaires controlling the economy. In many cases they write the legislation passed into law, laws giving legal status to corporate greed. Corporate profits, according to the most conservative supporters of capitalist theory, will "trickle down" to the vast majority who are not wealthy. In authoritarian countries, the government literally owns the economy outright. But in democratic countries powerful economic interests own the government. In the United States, we are rapidly approaching a political situation where a wealthy oligarchy composed of major corporate CEOs controls the House of Representatives and the Senate. The United States is in danger of becoming a fascist state as democratic process are hijacked by corporate greed.

If we are to somehow lower carbon and methane emissions flowing into the atmosphere, corporate ownership of politicians and governmental bureaucrats must end. But I do not mean that economic interests should not be allowed to influence political decisions. What I do mean to suggest is that corporate interests should not have the power to shove aside the wider communal interests of citizens, particularly in environmental issues affecting all citizens, rich and poor. Given the need to replace monstrosities like fossil fuels with renewable energy, which means putting the breaks not only on oil and coal abstraction through drilling and fracking but also on strip mining coal, "it is natural that we must," as Bill McKibben writes, "think big." He continues:

> Even for renewable energy, size makes a certain kind of sense: though the sun shines on every human, and wind rustles every blade of grass, it shines and rustles a good deal harder depending on your location. If you're going to build concentrated solar power arrays, start in the desert of the Southwest (or for Europe, in North Africa and the Iberian Peninsula); the biggest wind farms need the steady gusts of the Midwest, or the reliable onshore breezes on either coasts of the Atlantic [or either coasts of the Pacific]. We've recently begun to build some of this infrastructure, including the transmission lines necessary to connect them to places where most people live.[1]

But here's another hiccup. Constructing huge green-energy systems is like "buying organic food at the supermarket; it's an improvement, where its growth isn't soaked in pesticides, but that produce is still traveling an enormous distance along vulnerable supply lines."[2] That is to say, building

1. McKibben, *Eaarth*, 187.
2. Ibid.

national grids for the production of green energy will not necessarily build stronger local communities. Purchasing green energy will simply build the profits of the corporations who control green-energy companies. The upside of this reality is that an increasing number of people worldwide are beginning to understand the economic and ecological benefits not only of living green but also of green-energy systems. How this is to be accomplished on an international scale is both an economic and political decision. For the millions of us practicing a religious path, it will be necessary to socially engage economic and political leaders, to push them to promote living locally while not allowing green-energy corporations to control the political process as they line their own corporate pockets at the expense of the majority of human beings and other sentient beings. The focus of an international socially engaged interreligious dialogue should start here.

The reality is this: the majority of human beings now living on this planet practice some particular religious Way. All religious Ways call faithful persons to create communal social structures that exhibit justice and compassion. This is the foundation for a worldwide interreligious dialogue seeking ways to protect the environment and defending the poor by building networks of socially engaged faithful persons. This interreligious dialogue must include dialogue with the natural sciences as a third partner. This is so because the science underlying climate change must be understood before action can be undertaken to reverse its planetwide damage. The sheer gravity of the ecological crisis we are now experiencing demands that every human being on planet Earth begin seeking the common good founded on compassionate justice for all human beings along with all sentient beings with whom we must share the resources this planet.

Bibliography

Ali, Abdullah Yusuf. *The Holy Qur'an: English Translation of the Meanings and Commentary.* N.p.: King Fahd Complex for the Printing of the Holy Qur'an, 2002.

Armstrong, Karen. *The Case for God.* New York: Knopf, 2009.

Armstrong. Regis J. et al., eds. *Francis of Assisi: Early Documents.* Vol. 1, *The Saint.* New York: New City Press 2002.

Ashworth, William, and Charles E. Tuttle, eds. *Encyclopedia of Environmental Studies.* New ed. New York: Facts on File, 2001.

Austin, W. J., trans. *The Bezels of Wisdom,* by Ibn al'Abari. Classics of Western Spirituality. New York: Paulist, 1981.

Borenstein, Seth. "Scientists Enlist the Big Gun to Get Climate Action: Faith." *Daily Mail,* December 6, 2015. http://www.dailymail.co.uk/wires/ap/article-3348252/Scientists-enlist-big-gun-climate-action-Faith.html/.

Borg, Marcus. *Jesus: Uncovering the Life, Teachings, and Relevance of a Religious Revolutionary.* San Francisco, HarperOne, 2006.

Bonhoeffer, Dietrich. *Creation and Fall: A Theological Exposition of Genesis 1–3.* Edited by John W. de Gruchy. Translated by Douglas Stephen Bax. Dietrich Bonhoeffer Works 3. Minneapolis: Fortress, 2008.

Brooke, John Hedley. *Science and Religion: Some Historical Perspectives.* Cambridge History of Science. Cambridge: Cambridge University Press, 1991.

Buchanan, Mark. "Trapped in the Cult of the New Thing." In *Christianity Today,* September 6, 1999, 63–72.

California Office of Planning. http//www.opr.ca.gov/.

Chan, Wing-tzit. *A Sourcebook in Chinese Philosophy.* Princeton: Princeton University Press, 1963.

Clayton, Philip, and Justine Heinzekehr. *Organic Marxism: An Alternative to Capitalism and Ecological Catastrophe.* Toward Ecological Civilization Series. Claremont, CA: Process Century Press, 2014.

Climate Central. "How Hot Was Summer 2014?" Sustainability. *Scientific American.* http//www.scientificamerican/article/how-hot-was-summer-2014/

Conkin, Paul Keith. *The State of the Earth: Environmental Challenges on the Road to 2100.* Lexington: University Press of Kentucky, 2007.

Cobb, John B., Jr. *Beyond Dialogue: Toward a Mutual Transformation of Christianity and Buddhism.* Philadelphia: Fortress, 1982.

———. *Can Christ Become Good News Again?* St. Louis: Chalice, 1991.

———. "Introduction." In *Back to Darwin: A Richer Account of Evolution,* edited by John B. Cobb Jr., 1–18. Grand Rapids: Eerdmans, 2008.

———. *Is It Too Late? A Theology of Ecology.* Rev. ed. Denton, TX: Environmental Ethics Books, 1995.

———. *Matters of Life and Death.* Louisville: Westminster John Knox, 1991.

———. "Preface." In *For Our Common Home: Process-Relational Responses to "Laudato Si."* Edited by John B. Cobb, Jr. and Ignacio Castuera. Toward Ecological Civilization Series. Anoka, MN: Anoka Century, 2015.

———. *Reclaiming the Church.* Louisville: Westminster John Knox, 1997.

———. *Resistance: The New Role of Progressive Christians.* Louisville: Westminster John Knox, 2008.

———. *Spiritual Bankruptcy: A Prophetic Call to Action.* Nashville: Abingdon, 2010.

———. *Sustainability: Economics, Ecology, and Justice.* Ecology and Justice. Maryknoll, NY: Orbis, 1992.

———. *Sustaining the Common Good: A Christian Perspective on the Global Economy.* Cleveland: Pilgrim, 1994.

———. *Transforming Christianity and the World: A Way beyond Absolutism and Relativism.* Edited and introduction by Paul F. Knitter. Faith Meets Faith Series. Maryknoll, NY: Orbis, 1999.

Cornille, Catherine, ed. *Criteria of Discernment in Interreligious Dialogue.* Interreligious Dialogue Series 1. Eugene, OR: Cascade Books, 2009.

Daly, Herman E., and John B. Cobb Jr. *For the Common Good: Redirecting the Economy toward Community, the Environment, and a Sustainable Future.* 2nd ed. Updated and expanded. Boston: Beacon, 1994.

Douglas, Gordon, and Ward McAfee. "Consumerism." In *Resistance: The New Role of Progressive Christians,* 55–74. Louisville: Westminster John Knox, 2008.

———. "Poisonous Inequality." In *Resistance: The New Role of Progressive Christians,* 75–96. Louisville: Westminster John Knox, 2008.

Dunne, John S. *The Way of All the Earth: Experiments in Truth and Religion.* Notre Dame, IN: University of Notre Dame Press, 1978.

Francis, Pope. *Praise to You (Laudato Si): On Care for Our Common Home.* San Francisco: Ignatius, 2015.

Goodall, Dominic, ed. and trans. *The Hindu Scriptures.* Based on an anthology by R. C. Zaehner. Berkeley: University of California Press, 1996.

Gore, Al. *An Inconvenient Truth: The Planetary Emergency of Global Warming and What We Can Do about It.* New York: Rodale, 2006.

Gove, Philip Babcock et al., eds. *Webster's Third New International Dictionary of the English Language.* Springfield, MA: Merriam-Webster, 2002.

Hansen, James. "Timeline for Irreversible Climate Change." April 22, 2008. Policy Innovations. *Yale Global Online.* http://www.policyinnovations.org/ideas/commentary/data/000049/.

Hirsch, E. D., Jr. et al., eds. *The New Dictionary of Cultural Literacy.* Completely revised and updated 3rd ed. Boston: Houghton Mifflin, 2002.

Ingram, Paul O. *Buddhist-Christian Dialogue in An Age of Science.* Lanham, MD: Rowman & Littlefield, 2008.

———. *The Dharma of Faith: An Introduction to Classical Pure Land Buddhism.* Washington DC: University Press of America, 1977.

———. *Living without a Why: Mysticism, Pluralism, and the Way of Grace.* Eugene, OR: Cascade Books, 2014.

————. *Passing Over and Returning: A Pluralist Theology of Religions*. Eugene, OR: Cascade Books, 2013.

————. *Wrestling with the Ox: A Theology of Religious Experience*. 1997. Reprinted, Eugene, OR: Wipf & Stock, 2006.

Kahraman, Elias, and Ahmed Boig. *Environmentalism: Environmental Strategies and Environmental Sustainability*. Environmental Remediation Technologies, Regulations, and Safety Series New York: New Science, 2010.

King, Sallie B. "Conclusion: Buddhist-Social Activism." In *Engaged Buddhism: Buddhist Liberation Movements in Asia*, edited by Christopher S. Queen and Sallie B. King, 401–56. Albany: State University of New York Press, 1996.

————. "Thich Nhat Hanh and the Unified Buddhist Church of Vietnam: Nondualism in Action." In *Engaged Buddhism: Buddhist Liberation Movements in Asia*, edited by Christopher S. Queen and Sallie B. King, 321–63. Albany: State University of New York Press, 1996.

————. "Through the Eyes of Auchwitz and the Killing Fields: Mutual Learning between Engaged Buddhism and Liberation Theology." Paper presented at the Annual Meeting of the Society for Buddhist-Christian Studies. San Diego, CA. November 21, 2014.

Knitter, Paul F. "Towards a Liberation Theology of Religions." In *The Myth of Christian Uniqueness: Toward a Pluralist Theology of Religions*, edited by John Hick and Paul F. Knitter, 178–200. 1989. Reprinted, Eugene: OR: Wipf & Stock, 2005.

————. *Without Buddha I Could Not Be a Christian*. Oxford: One World, 2009.

Mason, Paul. "The End of Capitalism Has Begun." *Guardian*, July 17, 2015. http://www.theguardian.com/books/2015/jul/17/postcapitalism-end-of-capitalism-begun/.

McDaniel, Jay B. *With Roots and Wings: Christianity in an Age of Ecology and Dialogue*. Ecology and Justice. 1995. Reprinted, Eugene: OR: Wipf & Stock, 2009.

McFague, Sallie. *Life Abundant: Rethinking Theology and Economy for a Planet in Peril*. Minneapolis: Fortress, 2000.

McKibben, Bill, ed. *American Earth: Environmental Writing since Thoreau*. Library of America. New York: Penguin Putman, 2008.

————. *Eaarth: Making Life on a Tough New Planet*. New York: Times Books, 2011.

Nhất Hạnh, Thích. *Interbeing: Fourteen Guidelines for Engaged Buddhism*. Edited by Fred Eppsteiner. Berkeley: Parallax, 1987.

Oakman, Douglas E. *Jesus and the Peasants*. Matrix. Eugene, OR: Cascade Books, 2008.

————. *The Political Aims of Jesus*. Minneapolis: Fortress, 2012.

Popper, Ian L. et al., eds. *Environmental & Pollution Science*. 2nd ed. Amsterdam: Elsvier/Academic, 2006.

Pannikar, Raimundo. "Nine Ways not to Talk about God." *Crosscurrents* 47 (1997) 149–53.

Porete, Marguerite. *The Mirror of Simple Souls*. Translated by Ellen L. Babinsky. Classics of Western Spirituality. New York: Paulist, 1993.

Rawlings-Way, Olivia M. F. "Religious Interbeing: Buddhist Pluralism and Thich Nhat Hanh." PhD diss., University of Sidney, 2008.

Reuther, Rosemary Radford, ed. *Women Healing Earth: Third World Women on Ecology, Feminism, and Religion*. Ecology and Justice. Maryknoll, NY: Orbis, 1996.

Sanders, James A. *The Monotheizing Process: Its Origins and Development*. Eugene, OR: Cascade Books, 2014.

Schimmel, Annemarie. *The Mystical Dimensions of Islam*. Chapel Hill: University of North Carolina Press, 1975.

Schluter, Michael. "Is Capitalism Morally Bankrupt? Five Moral Flaws and Their Social Consequences." *Cambridge Papers* 18/3 (2009) 1–4.

Schumacher, E. F. *Small Is Beautiful: Economics as if People Mattered.* Harper Torchbooks. New York: Harper & Row, 1973.

Shuman, Michael H. *Going Local: Creating Self-Reliant Communities in a Global Age.* New York: Routledge: 2000.

Smith, Adam. *The Wealth of Nations.* Edited with an introduction by Kathryn Sutherland. Oxford World's Classics. Oxford: Oxford University Press, 2008.

Smith, Barbara Herrnstein. *Scandalous Knowledge: Science, Truth, and the Human.* Science and Cultural Theory. Durham: Duke University Press, 2005.

Streng, Frederick J. *Emptiness: A Study of Religious Meaning.* Nashville: Abingdon, 1967.

Unno, Taitesu. *River of Fire, River of Water: An Introduction to the Pure Land Tradition of Shin Buddhism.* New York: Doubleday, 1998.

———. *Shin Buddhism: Bits of Rubble Turned into Gold.* New York: Doubleday, 2002.

Vinot, H. D. *The Oxford Handbook of Hindu Economics and Business.* New York: Oxford University Press, 2012.

Voice of America. "2000 to 2010: A Period of 'Unprecedented' Weather Extremes, Scientists Say." July 12, 2013.

Waley, Arthur. *Three Ways of Thought in Ancient China*, 115–62. London: Allen & Unwin, 1939.

———. *The Way and Its Power: A Study of the Tao Te Ching and Its Place in Chinese History.* UNESCO Collection of Representative Works. Chinese Works. New York: Grove, 1958.

Watson, Burton, trans. *The Complete Works of Chuang Tzu.* UNESCO Collection of Representative Works. Chinese Works. New York: Columbia University Press, 1968.

Welch, Holmes. *Taoism: The Parting of the Way.* Rev. ed. Boston: Beacon, 1965.

Whitehead, Alfred North. *Process and Reality: Corrected Edition.* Edited by David Ray Griffin and Donald Sherburne. Gifford Lectures 1927–28. New York: Free Press, 1985.

World Meteorological Organization. http://learningenglish.voanews.com/content/a-period-of-weather-extremes/1699429.html/.

Name Index

Scripture Index